How to Write Technical Reports

LIVERPOOL JMU LIBRARY

3 1111 01392 0390

Lutz Hering · Heike Hering

How to Write Technical Reports

Understandable Structure, Good Design, Convincing Presentation

 Springer

Dr. Lutz Hering[†]
Am Ricklinger Holze 14
30966 Hemmingen
Germany

Dr. Heike Hering
Am Ricklinger Holze 14
30966 Hemmingen
Germany

ISBN 978-3-540-69928-6   e-ISBN 978-3-540-69929-3
DOI 10.1007/978-3-540-69929-3
Springer Heidelberg Dordrecht London New York

Library of Congress Control Number: 2010933599

© Springer-Verlag Berlin Heidelberg 2010
This work is subject to copyright. All rights are reserved, whether the whole or part of the material is concerned, specifically the rights of translation, reprinting, reuse of illustrations, recitation, broadcasting, reproduction on microfilm or in any other way, and storage in data banks. Duplication of this publication or parts thereof is permitted only under the provisions of the German Copyright Law of September 9, 1965, in its current version, and permission for use must always be obtained from Springer. Violations are liable to prosecution under the German Copyright Law.
The use of general descriptive names, registered names, trademarks, etc. in this publication does not imply, even in the absence of a specific statement, that such names are exempt from the relevant protective laws and regulations and therefore free for general use.

Cover design: WMXDesign GmbH, Heidelberg

Printed on acid-free paper

Springer is part of Springer Science+Business Media (www.springer.com)

Preface

Technical Reports are usually written according to general standards, corporate design standards of the current university or company, logical rules and practical experiences. These rules are not known well enough among engineers. There are many books that give general advice in writing. This book is specialised in how to write Technical Reports and addresses not only engineers, but also natural scientists, computer scientists, etc. It is based on the 6th edition published in 2008 by Vieweg in German and is now published as 1st edition by Springer in English.

Both authors of the German edition have long experience in educating engineers at the University of Applied Sciences Hannover. They have held many lectures where students had to write reports and took notes about all positive and negative examples that occurred in design reports, lab work reports, and in theses. Prof. Dr. Lutz Hering has worked for VOLKSWAGEN and DAIMLER and then changed to the University of Applied Sciences Hannover where he worked from 1974 until 2000. He held lectures on Technical Drawing, Construction and Design, CAD and Materials Science. Dr. Heike Hering worked nine years as a Technical Writer and was responsible for many CAD manuals in German and English. She is now employed at TÜV NORD Akademie, where she is responsible for E-Learning projects, technical documentation and software training and supervises students who are writing their theses. Prof. Dr.-Ing. Klaus-Geert Heyne joined the team as co-author for the 2nd German edition. He redesigned chapter 5 "Presenting the Technical Report". He contributes his experiences from Motorenwerke Mannheim AG (1978 to 1985) and University of Applied Sciences Wiesbaden from lectures about Combustion Engines, Technical Mechanics, and Technical Communication.

This book answers questions of engineering students and practitioners occurring when writing Technical Reports or preparing presentations on the PC. These questions refer to contents as well as formal aspects. Such questions occur during the whole work on the report or presentation from the beginning to the end. Therefore this book is designed as a guideline or manual „How to write Technical Reports". It is ordered by timeline along the process of writing Technical Reports into the three phases **planning, creation, and finishing**.

My father died in March 2004, Prof. Heyne prepares himself for retirement. I will continue this book as a guide with many examples and strong relationship to practical technical writing. Many comments of the German readers helped to improve this book. I hope that I will get similar positive feedback from international readers. If possible, please add example texts and figures, which I may publish in this book and correct menu translations, because I only have the Microsoft Office and Open Office programs in German. Please contact heike.hering@gmx.de.

Hannover, March 2010   *Heike Hering*

Contents

1 Introduction

People communicate in their spare time and in the professional area. They communicate either in oral or in written form. If they communicate about technical topics, this process is called technical communication. If they communicate in written form, they write or read "Technical Reports". If the Technical Report is communicated in oral form, it is a presentation to an audience.

ISO 5966 "Documentation – Presentation of scientific and technical reports" defines, that a **scientific or Technical Report describes a research process or research and development results or the current state-of-the-art in a certain field of science or technology**. Therefore all documents in the following list are Technical Reports, if they deal with a technical subject:

- reports about laboratory experiments
- construction and design reports
- reports about testing and measurements
- various theses written at the end of study courses, doctorate theses
- articles or reports about research works in scientific journals
- project reports etc.

A Technical Report can be defined as follows:

Technical Report = • report about technical subjects
  • written in the "language of science and technology" (special terms and phrases, display rules etc.)

In general, Technical Reports must comply with the following request:

☞ *Technical Reports must have a high level of systematic order, inner logic, consistency etc.*

The Technical Report shall bring **clarity** to the reader! This means, the reader must understand the topics described in the Technical Report in exactly the same manner as the author has meant it without any feedback or answers from the author. This can be checked as follows:

☞ *Imagine you are a reader who has basic technical knowledge, but no detailed knowledge about the topic or project described in the Technical Report. This fictive reader shall understand the Technical Report without any questions!*

This book is primarily addressed to readers with basic knowledge or people who are working in the various fields of engineering coming from universities and companies, i. e. it is primarily addressed to engineers and technicians, natural and computer scientists etc.

Today it is increasingly important to **present your ideas and work results** in Technical Reports to the scientific community, in interdisciplinary teams, to fund-

L. Hering, H. Hering, *How to Write Technical Reports*,
DOI 10.1007/978-3-540-69929-3_1, © Springer-Verlag Berlin Heidelberg 2010

ing organizations and the interested public **in a positive, professional manner**. However, this is sometimes very difficult for engineers and natural scientists. Too often they are not good sales people, in many cases they prefer to cope with technical problems. Yet, it is not all that difficult to present one's working results in a logical, clearly reproducible and interesting way to create the impression among your audience that this work was done by an experienced professional.

You can avoid mistakes and obstacles that other people – including the authors – have experienced before, if you read this book thoroughly or consult it when you have questions while preparing your next Technical Report.

It starts with taking a written report into your hands. Is it bound properly? Is it stored in a clean, tidy and wrinkle-free binder? Is there a clearly understandable title leaf? After you have got a rough overview of the contents you may ask: Does the title give sufficient and representative information about the contents of the Technical Report?

If you go into more detail, the following questions may occur. Is there a table of contents? Does it list page numbers? Is the table of contents ordered by logical rules, can you recognize the *"backbone"*? Does the report describe the starting point of the situation or project in an understandable way? Did the author critically reflect the task at the end of the report? Does the report contain citations? Is there a list of references etc.? Can you find tables, figures and references easily and are they designed according to common rules? If such formal requirements are not fulfilled, you will irritate your readers. Your readers will then have unnecessary difficulties in reading and understanding your message. This also influences how your project, your work results and you as a person are accepted.

For writing Technical Reports **word processors or desktop publishing systems** like **Microsoft Word**, **Open Office Writer,** etc. are used. At various spots in the text you will find hints, how to use Microsoft Word in an efficient, time-saving manner. If you use programs that are similar to Word, the program features will probably operate in a similar way. Hints how to use Open Office Writer are collected in a separate section. To create slide shows you will use **presentation programs**, such as **Microsoft PowerPoint**. Where it fits with the text and examples in this book, especially in chapter 5, you will find hints, how to create slides with Microsoft PowerPoint. Hints how to create slides with **Open Office Impress** can also be found in a separate section.

This book is designed to be lying beside the PC. Its layout uses little space to keep the production price low. However, it can be used as an example for creating your own Technical Reports. Terms from the fields documentation and printing technology can be found in appendix B "Glossary – terms of printing technology".

When working yourself though this book you can acquire the knowledge you need to write Technical Reports and presentations. **The concept of this book is that it shall answer questions instead of putting up new questions**. This book shall be a **guideline or manual how to write Technical Reports**. How is that meant? A user of a complicated technical product, like a video recorder, uses his instruction manual to be able to use the technical product. All functions of the product are described in detail in the instruction manual. The manual also lists all required warnings that allow safe usage of and working with the product.

Being an author, you can use this book similarly as an author's manual. In addition, you will get important information regarding how to avoid mistakes and obstacles during the presentation of your Technical Report. Moreover, this book will show you many important rules and checklists for text, table and image creation as well as for working with literature. Applying these rules and hints will make your Technical Reports readable and clearly understandable and comprehensible for your audience.

In accordance with the manual character of this book you – our audience – will often be personally addressed, so that the given information will reach you in an easy readable and motivating way. In doubt we used simple instead of complicated sentences to improve the understandability of the texts. Moreover we have kept several layout **rules**, which shall help you to orient yourself:

- Orders, notes, intermediate summaries etc. are written in italic letters and marked with a pointing hand: ☞.
- Series of menu commands are listed in their click sequence, separated by a dash, example: Format – Character.
- Graphics just illustrating the current text are used without a figure subheading.
- Examples are often indented.
- Important words are marked by boldface typing, so that you can find the required information quicker.
- The numbering of tables, figures and checklists, which also appear in the according list (of figures etc.) follows the syntax <chapter number>-<current number>. In examples the numbering syntax is <current number>.

If you read this book from the first to the last page you will notice, that **several information is presented more than once**. This was done on purpose. Most information required to create a Technical Report is closely linked with other pieces of information. In order **to present each section** of this book **as complete as possible** in itself and to avoid too many cross-references which would disturb fluent reading, we tried to give all the information you need to complete the task which is just described in the current section of the book.

☞ We *recommend all of you who are not very experienced in writing Technical Reports to read chapter 2 "Planning the Technical Report" and subchapter 3.7 "Using word processing and desktop publishing (DTP) systems", before writing your next Technical Report.*

Each writer's problem described in this book has occurred in Technical Reports submitted by students or during the authors supervised the writing of diploma, bachelor, master or doctorate theses. In addition the daily professional experience of the authors and many comments of our (German) readers have influenced the contents and layout of this book. Therefore this book reports **"from practical experience for practical usage"**.

2 Planning the Technical Report

Technical Reports shall be written so that they reach your readers. This requires a high level of systematic order, logic and clarity. These understandability aspects must already be taken into account, when you plan the necessary work steps. This is the only way to perform all work steps accurately. As a result all facts about the described items or processes and the thoughts of the writer of a Technical Report become clear for the reader without any questions and without doubt.

In technical study courses a systematic approach is used to solve tasks and larger projects. Tasks are solved in the sequence *planning, realization* and *checking*. This approved approach should be applied in a similar way when creating Technical Reports. Here the necessary work steps can be grouped in the phases *planning, creation and finishing (with check-ups)*. However, before describing the single measures in the planning process we will present a general overview of all required work steps to create a Technical Report.

2.1 General overview of all required work steps

The following **Checklist 2-1** shows all required work steps.

Checklist 2-1   Required work steps to create Technical Reports

- Accept and analyze the task
- Check or create the title
- Design a 4-point-structure
- Design a 10-point structure
- Search, read and cite literature
- Elaborate the text (on a computer)
- Create or select figures and tables
- Develop the detailed structure
- Perform the final check
- Print copy originals or create PDF file
- Copy and bind the report
- Distribute the report to the defined recipients

Work steps
to be performed
partly parallel
or overlapping

This list is complete, but the clarity can be further improved. To accomplish this, network planning is applied.

L. Hering, H. Hering, *How to Write Technical Reports*,
DOI 10.1007/978-3-540-69929-3_2, © Springer-Verlag Berlin Heidelberg 2010

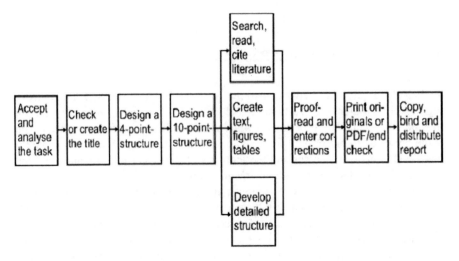

This network plan is always repeated when the different steps to create a Technical Report are described, where the current work step is marked in gray.

Please keep in mind, that the amount of work to create a Technical Report is regularly *completely* underestimated. To avoid this, make a proper assumption of the required time and double the estimated timeframe! Start early enough to create your Technical Report – no later than after 1/3 of the total timeframe of your project.

2.2 Accepting and analyzing the task

When you write a Technical Report, there is nearly always a task, which you either selected yourself or it was defined by someone else. You should analyze this task precisely during the planning of the Technical Report, **Checklist 2-2**.

Checklist 2-2   Analysis of the task to write a Technical Report

- Who has defined the task?
  - a professor or an assistant (in case of a report written during your studies)
  - a supervisor
  - the development team
  - a consulting company
  - a customer
  - you yourself (e. g. if you write an article for a scientific journal)
- Did I understand the task correctly?
- Who belongs to the target group? For whom do I write the report?
  Please take notes accordingly!
- Which contents shall my report contain? Please write that down!

- Does the task already contain a correct and complete title?
- Which work steps are necessary?
- Which help and assistance do I need?
  - help by people, e. g. *advice-giving specialists*
  - help by equipment, e. g. *a color laser printer*
  - help by information, e. g. *scientific literature*

This work step is called "Accept and analyze the task" in the network plan it is marked in gray.

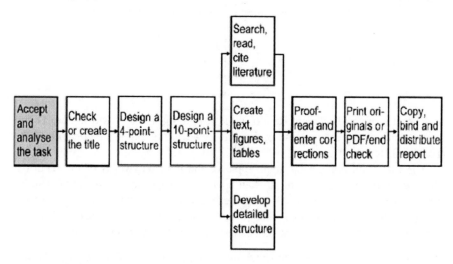

In addition, during the planning of the report the following questions must be answered:

- Which shall be the title of the report?
  (develop a proposal and discuss it with the supervisor or customer)
- Which work steps that are not mentioned in the network plan need to be accomplished?
- Which background knowledge, interests and expectations do the readers of the Technical Report have?
- How do I organize the required help?
- Which help and work steps are time-critical?

2.3 Checking or creating the title

In the next step, see network plan, the title which in most cases is predefined by the supervisor or customer must be checked and evtl. a new title must be created.

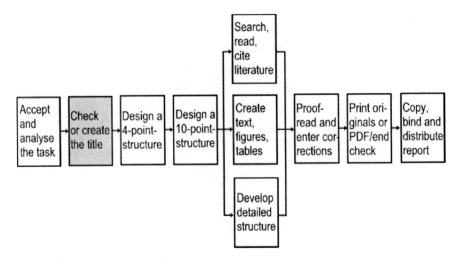

The title of the Technical Report is the first thing a reader will notice. Therefore it shall create interest and curiosity to learn more about the contents of the Technical Report.

The title shall contain the main topic or the main keywords of the report, it shall be short, precise and true. It shall have a good speech melody and create interest. Explaining or additional aspects can appear in a subtitle. In any case the title (and subtitle if applicable) shall describe the contents of the Technical Report accurately and it must not create undesired associations or wrong expectations.

☞ *These demands, the title of a Technical Report must fulfil, must also be fulfilled by all other titles and headings of paragraphs, figures, tables etc.*

In many cases the task can already be used as the title of the Technical Report. Here are some examples of such tasks:

- Design of a drilling rig
- Outline of a sprayer shredding rig
- Analysis of component combinations for sales optimization
- Equipment of a meeting room with radio technology

Even, if a title seems to be usable, we recommend that you systematically create possible title variants. Then you (and eventually the supervisor or customer) can decide which title shall be used. It is also possible to use the task as a working title in the beginning of your project.

The final decision which title shall be used can then be found later during your project without time pressure. The following **Checklist 2-3** shows again all requirements of the title of the Technical Report as a conclusion.

Checklist 2-3   Requirements of the title of the Technical Report

- The title must be clear, true, honest, short and accurate,
- it must contain the main topics or main keywords (for data base searches!),
- it must create interest and curiosity,
- have a good speech melody and
- eventually an additional subtitle.

☞ *Write down the main keywords which characterize your Technical Report by hand, connect these keywords to a title, create several title variants by using different keywords and select the "best" title.*

Now the process to create a title will be explained in an example.

Example for the creation of a title

We are looking for the title of a doctorate thesis. In the doctorate project a computer program has been developed, that allows the selection of the materials of designed parts depending on the stress on the part, abrasion requirements etc. The designer enters the requirements which the material must fulfill and the system provides the materials, which are stored in its database and match the given requirements. It has been quite early in the project that the developer of the system, the doctorate candidate, has defined the term "CAMS" = Computer Aided Material Selection to describe the purpose of the program.

The doctorate candidate starts to create a title for his thesis as described above. He starts to **write down the keywords** that shall be contained in the title.

- material selection
- design
- education
- CAMS
- with computer

The next step is to **combine the keywords** to get different titles:

Contribution to computer-aided material selection
Computer-aided material selection in design
Computer-aided material selection in design education
Computer Aided Material Selection = CAMS
CAMS in design education
Help to select materials by the computer
Computer application for material selection
CAMS in design
Design with CAMS
Computer support in design education
Material selection with the computer

Since the doctorate candidate has defined the term CAMS it shall definitely appear in the title of his doctorate thesis, so he decides to select the following title:

Computer Aided Material Selection – CAMS in Design Education

The following list of work steps summarizes the process to find a good title for your Technical Report.

☞ *Use the following work steps to create the title:*
  • *write down the task*
  • *write down the keywords which characterize the report*
  • *combine the keywords to a title*
  • *find new titles by varying the usage of these keywords*
  • *read possible titles aloud to optimize the speech melody*
  • *select the "best" title*

After the title has been created, the next step is to **design the structure**.

2.4 The structure as the "backbone" of the Technical Report

In our network plan to create Technical Reports we have now arrived at the two last work steps in the phase of planning the report. These work steps are designing the 4-point- and 10-point-structure.

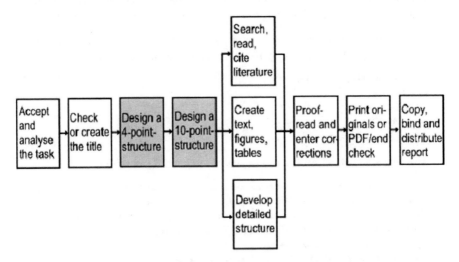

Since **designing the structure is the main step of planning** the Technical Report, we want to give you an introduction to the underlying rules of logic and formal design in the next three sections. Then we will show you how to create a logical structure and provide you with four structure patterns in sections 2.4.4 and 2.4.5. Many people do not distinguish properly between the terms "structure" and "table of contents" (ToC). Therefore we define these terms as follows:

Structure:	*without* page numbers,	contains the logic,	is **intermediate result**;
ToC:	*with* page numbers,	allows searching,	is **final result**.

The typographic design or layout of the structure or table of contents is not a work step in the phase of planning the Technical Report, but it belongs to creating the Technical Report. Therefore it is described in section 3.1.2.

2.4.1 General information about structure and table of contents

The **structure** (while writing the Technical Report) or the table of contents (after finishing the Technical Report) is the **"front entrance door"** into your Technical Report. It is the next piece after title leaf and Preface/Foreword and/or Summary that is read in larger documents like books, applications for research projects, final reports of research projects, design descriptions, etc.

☞ *A good structure is so important for the understandability and plausibility of texts – even of short texts like e-mails –, that you should always structure every text that exceeds the amount of about one page with intermediate headings – at least every text describing facts.*

The structure allows you to get a **quick overview**
• to find your way into the contents of the Technical Report,
• to get help from your supervisor, and
• to evaluate/grade your Technical Report.

☞ *Therefore you should always take the current state of the structure with you when you are going to discuss the current status of your project with your supervisor (boss, assistant, professor, etc.) or with your customer. They ask for it quite frequently!*

Other materials which are not necessarily required (e. g. literature references and copies which are important or difficult to get) should also be available in the meeting.

☞ *For each reader of a Technical Report the structure is the most important tool to understand the contents. Therefore you should not make any compromises with yourself when designing the structure! This also holds true for writing the whole Technical Report. Wherever you are not confident with your report, the supervisor will criticize this not so successful part of the Technical Report in most cases – and a customer will make up his/her mind.*

The information which forms your Technical Report will only be sorted into the drawers which are defined by the structure. Thus, creating the structure is the creative part of the work. Writing the text is just "craftsmanship", which requires only routine.

2.4.2 Rules for the structure in ISO 2145

When explaining the term **structure**, it is also necessary to discuss **levels of document part headings**. People use terms like chapter, subchapter, section, subsection, main item, item, clause, sub clause, paragraph, listing, etc. To refer to document parts of various levels, but these terms are not used by all people in the same sense.

If you look into the standard ISO 2145 "Documentation – Numbering of divisions and subdivisions in written text", you will find that the standard uses the terms "main divisions" for the 1st level, "subdivisions" for the 2nd level and "further levels of subdivision" for the 3rd and all lower levels.

However, this terminology does not comply with the general usage of language of most people, who think of large documents being subdivided into chapters. Therefore in this book we will use the following system of terminology based on the term "chapter".

title (whole report)
   chapter
      subchapter
         section   —   *document part headings*
            subsection

This gives clarity and the text blocks on all different levels of hierarchy can be individually named.

To continue this hierarchy, the following terms should be used to refer to **text elements**.

> *paragraph*
>   *sentence*
>   *word*
>   *character*

Apart from text the document parts can also contain other objects that illustrate the statements or messages given in the text. In many texts the following objects which are equivalent to paragraphs occur.

> *table*
> *figure*
> *equation*
> *list*

The standard ISO 2145 "Documentation – Numbering of divisions and subdivisions in written text" is the most important standard for creating the structure of a document. It is relevant for all types of contents, i. e. for texts dealing with technology, commerce, humanities, laws, medicine etc. and for all kinds of written documents like manuscripts, printed works, books, journal articles, manuals, directions for use and standards.

The standard itself has the following structure:

- Scope and field of application
- Numbering of divisions and subdivisions
- Citation of division and subdivision numbers in text
- Spoken form

The numbering of document parts is in consecutive Arabic numerals. Each document part can be further subdivided into at least two subdivisions. The subdivisions are also continuously numbered. The document part hierarchy is expressed by a **full stop between the numbers of subdivisions on different levels**. No full stop shall be used at the end of the final level, i. e. chapter numbers will not have a full stop at the end.

According to ISO 2145 there can be any number of document part levels, but the number should be limited, so that reference numbers are still easy to identify, to read and to cite. We recommend, that the **number of document part levels should be limited to three, if possible.** Example: A document has nine chapters numbered 1, 2, 3, etc. Chapter two is e. g. subdivided into subchapters 2.1 and 2.2. Subchapter 2.1 is subdivided into sections 2.1.1, 2.1.2 and 2.1.3. To keep the document numbers simple, we recommend, that the **number of equal document parts on the same level should not exceed nine**.

A document part number "0" (zero) can be assigned to the first division of each level, if the contents of that document part has the character of a foreword, preface, introduction or similar type. This is more frequently used for the chapter and subchapter levels (numbers 0 and n.0) than for the further levels.

Technical reports require a high level of tidiness and logic. This logic must naturally speaking be reflected in the structure. Therefore, when writing a Technical Report, the author must always **keep the inner logic from the first sketch to the final version of the structure**. The sequence of work steps described in 2.4.4 "Work steps to create a structure and example structures" will nearly automatically result in a good and logical structure. Before this is described in detail, we want to introduce important rules for document part numbers and document part headings, because these rules will be applied in section 2.4.4 during the creation of structures.

2.4.3 Logic and formal design of document part headings

Document part numbers and document part headings express the logic of the sequence of thoughts and work steps (the "thread" or "backbone") in the Technical Report. For many people "logic" has something to do with mathematics and its rules. But there is also the logic of language, which is examined in many intelligence tests beside the mathematical logic.

☞ *You should be able to optimize your own structures according to the logical sequence of thoughts and work steps described in your Technical Report. This requires that you develop the ability to check your own structures for proper logic of language.*

This recommendation will now be explained by means of examples and further descriptions. It is a key requirement of a logical structure that **different document part headings on the same level of hierarchy must be equally important and consistent**. Therefore the following part of a structure is **not logical**:

3.5 Technical evaluation of concept variants
     3.5.1 Technical evaluation table
3.6 Economical evaluation table

It happens quite frequently in Technical Reports and other larger documents or books that a document part heading is **subdivided only once**. However, this is not logical, because the subdivision into document parts of a lower than the current level happens, because *several* aspects of a superordinated topic shall be distinguished from each other.

Therefore it is **not logical, to subdivide** a higher-level topic **in the next lower document hierarchy level into only one document part heading**. Here you should either add one or more additional document part headings of the same hierarchy level or leave the superordinated topic without subdivision. Here is a correct alternative for the bad example above:

3.5 Technical-economical evaluation of the concept variants
     3.5.1 Technical evaluation of the concept variants
     3.5.2 Economical evaluation of the concept variants
     3.5.3 Summarizing evaluation of the concept variants in the s-diagram

Here is another example.

Not logical:

1 Introduction
   1.1 Starting point
2 Basics of metal powder production

Logical solution:

1 Introduction
   1.1 Starting point
   1.2 Goals of this work
2 Basics of metal powder production

Other logical solution:

1 Introduction
2 Basics of metal powder production

Each document part heading shall be complete in itself and represent the contents of the document part properly! It shall be short, clear and accurate as the title of the whole Technical Report. **Document part headings that consist of one word only can often be improved**. Exceptions from this rule are *generally-used* single words like Introduction, References, Appendices etc.

Please find a summary of the rules mentioned so far plus additional rules for document part headings and numbers in the following **Checklist 2-4**.

Checklist 2-4   Rules for document part numbers and headings

Rules of logic
- Full stops in section numbers define the hierarchy level in the document
- Document part numbers0, n.0 etc. can be used for foreword/preface, introduction etc.
- Each hierarchy level consists of at least two document parts which are logically of equal importance
- The document part heading may not be the first part of the first sentence of the first paragraph in the appertaining text, but it must be an own and independent element of the Technical Report. The first sentence of the following text must be a complete sentence, which may pick up or repeat the contents of the document part heading.

Formal rules
- The declaration in lieu of an oath, task, abstract, foreword/preface and table of contents always get a document part heading, but no document part number.
- At the end of document part number and document part heading *never* use a punctuation mark like period, colon, question mark, exclamation mark etc.
- It is unusual to formulate the document part heading as a complete sentence or as a main clause with one or more subclasses.
- At the end of document part headings there is *never* a reference to the literature like "[13]".

Layout rules
- If you want to create the table of contents automatically with your word-processing program, use the standard format patterns or formatting styles resp. in the continuous text. Format chapter headings with "Heading 1", subchapter headings with "Heading 2", section headings with "Heading 3" etc. You may as well change the formatting of these format patterns to modify the appearance of the headings in the continuous text. To modify the appearance of the table of contents, change the format patterns "ToC 1", „ToC 2" or how ever they are called in your word processor, see also 3.7.4. It is general use, that the document part headings appear in boldface typing and larger than the normal text. They must not be underlined.
- Please avoid capital letters in headings and table items (in the table of contents, list of figures, list of tables etc.), because this is substantially more difficult to read than the ordinary mixture of capital and small letters.
- It is not clearly defined in ISO 5966 "Documentation – Presentation of scientific and technical reports" and in other documentation standards (e. g. ISO 8 "Documentation - Presentation of periodicals"), which distance document part headings should have from the previous and following text. In the different standards this distance is sometimes alike (ISO 5966) and sometimes the distance to the previous text is larger (ISO 8). If the distance above a document part heading is larger than the distance below, it becomes clearer, which heading belongs to which text, and therefore we recommend this layout principle.

The rules above hold similarly true for titles of tables and figures/illustrations with the following exceptions:

- At the end of table and figure titles *there must appear* a citation, if the figure or table is created by other authors.
- There are other rules for table numbers and figure numbers than for document part numbers. Figures and tables are either chronologically numbered through the complete Technical Report or the numbers are combined using the chapter number and a running number within the current chapter. Often these two components of the table or figure number are connected by a hyphen, see also 3.3.2 and 3.4.2.
- If the list of figures and list of tables shall be created automatically from the figure and table titles, you must not use manual formatting to influence the appearance of the text, but you should apply appropriate format patterns or formatting styles resp., see also 3.3.2 and 3.4.2 as well as 3.7.4.

After we have introduced you to the most important rules for the formulation and layout of document part numbers and headings, now we can use that knowledge to create the structure.

2.4.4 Work steps to create a structure and example structures

The creation of the structure should be divided into several consecutive work steps. Starting from the working title (or the final title) the **main topic or core message** of the Technical Report should be formulated **in *one* sentence**. This information will **then be further subdivided** into document part headings up to the complete final structure which will appear in the Technical Report as the table of contents later. To develop the final, logical structure, the following procedure has quite frequently been successfully applied, **Checklist 2-5**.

Checklist 2-5   Work steps to create a structure

1. Formulate the title of the main topic, main target or core message of the Technical Report in one sentence.
2. Subdivision into 3 to 4 main items (4-point-structure).
3. Further subdivision into 8 to 10 main items (10-point-structure).
4. Further subdivision of extensive main items.
5. Further subdivision into the final detailed structure parallel with the further elaboration of the Technical Report.
6. Last but not least: Check whether the document part numbers and headings are identical in the structure and in the text (check for completeness and correctness) and add page numbers to the structure to make it a table of contents, if the table of contents shall not be automatically created by your word processor.

If you apply this procedure, the logical order of information which is already defined in the 4-point-structure cannot be lost any more up to the final detailed structure, when you add divisions or split divisions into subdivisions!

Now this **procedure shall be explained by means of examples**. The examples are derived from a report about the enhancement of the computer network at a customer company (a project report), a design report, a report about executed measurements (laboratory report), and a diploma thesis, where a computer program has been developed. Naturally speaking, this procedure can be applied for any other type of report like for literature research works etc. **Checklist 2-6** provides a summary.

Example 1: Report about the enhancements of a computer network

Title of the report:
Equipment of a meeting room with radio technology

1st Step: Formulate the main topic (main target) of the Technical Report
The computer network in the customer company shall be enhanced so that there are two additional internet access points for external staff members in the training room and two additional internet access points for training participants in the lounge.

2nd Step: Subdivision into 3 to 4 items (4-point-structure)
- Analysis of the customer's requirements
- Planning of the new network structure
- Realization of the network enhancements in the customer company
- Billing and payment

3rd Step: Subdivision into 8 to 10 items (10-point structure)
1 Introduction
2 Analysis of the customer's requirements
3 Planning of the new network structure
4 Preparing work steps
5 Realization of the network enhancements in the customer company
6 Inspection
7 Billing and payment
8 Conclusions

4th Step: Further subdivision of extensive main items
Chapter 2 can be subdivided into the steps status quo-analysis and target situation-analysis. Chapters 3, 4 and 5 and 9 Appendices have also been subdivided further in the original work.

5th Step: Further subdivision into the final detailed structure
                parallel with the further elaboration of the Technical Report
3 Planning of the new network structure
   3.1 Collection of offers from hardware suppliers
   3.2 Benefit analysis and decision of suppliers for the hardware to be used
   3.3 Planning the wiring
   3.4 Planning of external services

Example 2: Design report

Title of the report:
Redesign of a production plant for Magnesium-Lithium-Hydrogen alloys

1ˢᵗ Step: Formulate main topic (main target) of the Technical Report
Weaknesses of the existing founding plant shall be improved by the redesign.

2ⁿᵈ Step: Subdivision into 3 to 4 main items (4-point-structure)
- State of the art
- Description of the existing weaknesses
- Description of the modifications

3ʳᵈ Step: Subdivision into 8 to 10 main items (10-point-structure)
1 Introduction
2 State of the art
3 Necessary modifications of the existing plant
4 Requirements the new plant shall fulfill
5 Redesign and reconstruction of the existing plant
6 Practical testing of the new plant
7 Evaluation of the tests with the new founding plant
8 Conclusions and outlook

4ᵗʰ Step: Further subdivision of extensive main items
Chapter 3 can be subdivided into the necessary modifications (possible usage of the plant for other technological processes, facilitated usage and handling, facilitated cleaning, improved safety while working with hydrogen etc.).
Chapter 5 can be subdivided into basic design principles applied for the redesign of the founding plant and design details.

5ᵗʰ Step: Further subdivision into the final detailed structure
              parallel with the further elaboration of the Technical Report
5 Redesign of the new plant
    5.1 Basic design principles and principle drawing
    5.2 Design details to realize the required modifications
        5.2.1 Basic design of the founding plant
        5.2.2 Temperature flow in the plant components
        5.2.3 Gas flow of inert gas and alloy gas
        5.2.4 Modifications of the casting device
        5.2.5 Flexible structure of the cast container via plugging system
        5.2.6 Inert gas container for the die-cast
        5.2.7 Central plant control via the control panel

Example 3: Report about executed measurements

Title of the report:
Damage detection with holographic interferometry

1ˢᵗ Step: Formulate main topic (main target) of the Technical Report
The deformation of a steel container under inner pressure shall be measured with holographic interferometry (target: identification of the influence of container geometry, welding zone, heat affected zone and intentionally added material flaws on the deformation of the steel container).

2ⁿᵈ Step: Subdivision into 3 to 4 main items (4-point-structure)
- State of the art
- Testing plant design
- Test execution
- Test results

3ʳᵈ Step: Subdivision into 8 to 10 main items (10-point-structure)
1 Introduction
2 State of the art
3 Testing plant design
4 Test preparation
5 Test execution
6 Evaluation of the interferograms
7 Estimation and classification of measurement flaws
8 Proposals for continuing works
9 Conclusions

4ᵗʰ Step: Further subdivision of extensive main items
Chapter 5 can be subdivided by the type of the executed work steps into estimation of the required inner pressure in the testing container, description of the unintended material flaws in the welding zone, the heat affected zone and the intentionally added material flaws, description of the measurement points, influence of the container geometry.
In chapter 6 the evaluation of the measuring results can be subdivided into the local influence of the weld seam and welding zone and the types of intentionally added material flaws.

5ᵗʰ Step: Further subdivision into the final detailed structure
                 parallel with the further elaboration of the Technical Report
6 Evaluation of the interferograms
    6.1 Relative deformation extremae
    6.2 Influence of the heat affected zone
    6.3 Influence of the welding bead
    6.4 Influence of the intentionally added material flaws

Example 4: Report about the development of software

Title of the report:
Computer-aided analysis and optimization of the understandability of technical texts

1ˢᵗ Step: Formulate the main topic (main target) of the Technical Report
Starting from existing approaches to improve the understandability of texts an interactive computer program shall be developed that measures the understandability of text and that stepwise improves the understandability of the text in constant dialogue with the user.

2ⁿᵈ Step: Subdivision into 3 to 4 main items (4-point-structure)
- Approaches to measure and improve the understandability of texts
- Development of the understandability improvement concept of docutune
- The program system docutune
- Documentation of the source code

3ʳᵈ Step: Subdivision into 8 to 10 main items (10-point-structure)
1   Introduction
2   Approaches to measure and improve the understandability of texts
3   Development of the understandability improvement concept of docutune
4   The program system docutune
5   Documentation of the source code
6   The practical use of docutune
7   Further development of docutune
8   Conclusions and outlook

4ᵗʰ Step: Further subdivision of extensive main topics
Chapter 2 deals with the state-of-the-art as it is described in the literature. It has been further subdivided into scientific approaches to the research on understandability, practically-oriented approaches to improve the understandability and Hamburg concept of understandability.

5ᵗʰ Step: Further subdivision into the final detailed structure
parallel with the further elaboration of the Technical Report
4   The program system docutune
   4.1   The menu structure of docutune
   4.2   The sequence of docutune's feature groups
      4.2.1   General overview of the sequence of all feature groups
      4.2.2   Sequence of the feature group Typography
      4.2.3   Sequence of the feature group Clarity
      ... (further sections for Logic, Shortness, Motivators)
      4.2.8   Sequence of the feature group Orthography
   4.3   Help features for searching and classification
      4.3.1   The word classification object
      4.3.2   The dictionary object

Checklist 2-6   Rules and tips for creating the structure

1st Step: Formulate main topic (main target) of the Technical Report
Here you should formulate the target of the project, the literature research, the
tests, the measurements, the design, the expert opinion or the report in general.
Even if it seems hard to accomplish: Write that down in one sentence only!
2nd Step: Subdivision into 3 to 4 main items (4-point-structure)
Examples:
- "Starting situation – Own contribution – Improvements of the situation –
  Summary"
- "State-of-the-art – Testing rig design – Test execution –
  Test results – Conclusions"

If you integrate the task of your project into your Technical Report as an independent chapter, then the chapter "Task" and the various chapters about fulfil-
ling the task (general draft, detailed design, computation of loads or testing rig
design, test execution, test results etc.) are each an individual chapter.
3rd Step: Subdivision into 8 to 10 main items (10-point-structure)
Possible structuring principles for the 3rd, 4th and 5th step are:
- by time sequence or
- by starting point conditions
- by project targets
- by possible alternatives
- by components or part groups
- by improvement steps
- by related topics or in the special case depending on the task

4th Step: Further subdivision of extensive main items
Possible structuring principles have already been mentioned in the 3rd step. We
recommend, that *before* writing the text for a chapter, you should create a temporary structure of this chapter into subchapters. In the same way, *before* writing the text for a subchapter, you should create a temporary structure of this
subchapter into sections or consciously decide that no further subdivision is
necessary etc. This recommendation corresponds to the sequence of work steps
to create a temporary 4-point- and 10-point-structure of the Technical Report,
before you start at all with writing text, searching for literature and collecting
other materials.
To reach your target group you should use common document part headings,
which your reader expects to find in your Technical Report. In a report about
laboratory experiments, a reader would for example expect document part
headings like testing rig design, test execution and test evaluation or test results. Therefore you should use these document part headings in your Technical Report and in your literature and material collection.
5th Step: Further subdivision into the final detailed structure
          parallel with the further elaboration of the Technical Report
This step needs no further explanation.

2.4.5 General structure patterns for Technical Reports

In the following we show you **structure patterns for often written types of Technical Reports**, which have been successfully used in practice. If you use such a structure pattern, you don't need to create a 4-point- and 10-point-structure.

At first we provide a structure pattern for a rough design description in which after analyzing the sub functions and the design solutions of the sub functions several concept variants are defined. These will then be evaluated according to the VDI guidelines for design methodology VDI 2222 and 2225 (see also 3.3.3 "The morphological box – a *special* table").

Structure pattern of a rough design description – several concept variants
1   Starting situation
2   Task
2.1   Task definition
2.2   List of requirements
3   Function analysis
3.1   Formulating the overall function
3.2   Subdivision into sub functions
3.3   Morphological box
3.4   Definition of the concept variants
3.5   Technical evaluation of the concept variants
3.6   Economical evaluation of the concept variants
3.7   Selection of the most useful concept variant with the s-diagram
4   Design
4.1   Design description
4.2   Computation of loads
5   Summary and conclusions
6   References
A   Bill of materials
B   Manufacturer documents

If you do not add manufacturer documents, just use the appendix "A Bill of materials". If you want to add printouts or plots or photocopies of technical drawings in reduced size, you can structure the appendices as follows: A Bill of materials, B Assembly drawing, C Component drawings, D Manufacturer documents. The bill of materials is actually not a part of the report, but it belongs to the set of drawings. Since in universities the drawings are transported in drawing rolls, it has proven to be practical, to add the bill of materials twice to the Technical Report: one copy of the bill of materials is added as an appendix of the bound Technical Report and the other is added to the set of drawings in the drawing roll.

If in industry during a presentation plotted drawings are fixed to the walls of the meeting room, the (enlarged!) bill of materials can also be hung up at the wall.

Please decide, whether you want to add a photocopy of the assembly drawing in reduced size to your Technical Report. It can either be added to an appendix (directly behind the bill of materials) or used in a text chapter (preferably in the chapter "Design description". If the photocopy of the assembly drawing in reduced size is bound in your Technical Report, this has another advantage: In the design description the parts can be referred to with their names and in addition with their position numbers, e. g. "Handle (23)" in the assembly drawing and in the bill of materials. However, when the first part name with added position number occurs in the text of the design description, you should explain, that the number is a position number and refers to the assembly drawing and the bill of materials. Identical parts, which are used in two different components (e. g. headless screws at two halves of a clutch), occur only once in the bill of materials in a common line with a common position number. The column "Number" contains the information, how often the part is used in the complete assembly. So the person who mounts the part can check, whether all required single parts are available.

Now we want to look at the structure patterns again. In the following you will find a structure pattern for a rough design description where the most useful design solutions of the sub functions are combined to only one concept variant (see also 3.3.3 "The morphological box – a *special* table").

Structure pattern of a rough design description – one concept variant

1   Introduction
2   Task
    2.1   Task definition
    2.2   List of requirements
3   Function analysis
    3.1   Formulating the overall function
    3.2   Subdivision into sub functions
    3.3   Morphological box
    3.4   Verbal evaluation of the design alternatives for the sub functions
    3.5   Description of the concept variant
4   Design
    4.1   Design description
    4.2   Computation of loads
5   Summary and conclusions
6   References
A   Bill of materials
B   Manufacturer documents

Now we want to give you a structure pattern for projects dealing with *laboratory experiments or other experimental works*. First there is an important rule:

Laboratory experiments must always be documented "reproducible"!

This means, that all information must be provided in detail so that another team of researchers can execute the experiments again under exactly the same conditions and they will get the same results.

Therefore the following information may never be left out:

- testing machine, device, or rig with manufacturer, type number and/or name, inventory number etc.
- all parameters set or selected at the machine, device or rig
- all measuring instruments, always with manufacturer, type number and/or name, inventory number, set or selected parameters etc.
- tested specimens with all required data according to the appertaining standard, regulation or guideline (ISO, EN, DIN or other), taken samples
- in experiments which are not standardized similar data regarding specimen shape, experiment parameters, temperatures, physical/chemical properties etc.
- all measured values or test results with all parameters
- used evaluation formulas with complete bibliographical data of used references

☞ *Provide so much information, that someone else will measure the same values or find out the same test results as you, if he/she executes the experiments exactly under the described conditions.*

Structure pattern of an experimental work

1 Target and scope of the test
2 Theoretical basics
3 The laboratory experiment/test
   3.1 Testing rig design
      3.1.1 Testing machine, plant, rig or device
      3.1.2 Used measuring instruments
   3.2 Test preparations
      3.2.1 Specimen preparations
      3.2.2 Setup of the starting conditions
   3.3 Test execution
      3.3.1 Execution of the preparation tests
      3.3.2 Execution of the main tests
   3.4 Test results
   3.5 Test evaluation
   3.6 Estimation of measurement flaws
4 Critical discussion of the laboratory experiments/tests
5 Conclusions
6 References
A Measurement protocols of the preparation tests
B Measurement protocols of the main tests

Now we want to give you a short introduction to a different type of document or Technical Report before we will provide you with a structure pattern for that document type.

Manuals and instructions for the usage for complex technical products should be written by technical writers. However, in practice they are often written by engineers. During engineering study courses the supervisor gives the task to develop a new electronic circuit or the design of a plant or rig. As part of the project it is always necessary to prepare the technical documents, but in more and more cases it is also required that the instruction manual for a potential user must be written. Therefore we will give you a structure pattern for manuals and instructions for use. Manuals and instructions for use are structured according to different schemes. They can be subdivided and numbered according to ISO 2145, but they can as well have document part headings without document part numbers, which are just layouted in boldface typing.

To provide more uniformity here, EN 62079 "Creation of instructions; Structure, contents and presentation" has been published. Among other information this standard describes, which information shall be given in which sequence in instruction manuals. Other definitions used in the following structure pattern are derived from DIN 31051 "Grundlagen der Instandhaltung (Basics of maintenance)", DIN 32541 "Betreiben von Maschinen und vergleichbaren technischen Arbeitsmitteln – Begriffe für Tätigkeiten (Running machines and comparable technical devices – Terms for work steps", VDI guideline 4500 "Technische Dokumentation (Technical documentation)".

☞ The information can either be presented according to the structure and logic of the product (product-oriented) or according to the sequence and logic of work steps during product usage (task-oriented).

Structure pattern for manuals and instructions for use

1 Before operating the machine/device
   1.1   Important information about the machine/device
          (Definition/description of the machine/device, description of the benefits, safety notes and warnings, overview of the functions)
   1.2   Supplied/delivered scope and optional parts
   1.3   Usage of the machine/device (Rules and regulations, safety notes and warnings, intended usage, unintended usage, documentation provided by third parties)
   1.4   Transportation of the machine/device
   1.5   Requirements regarding the site
   1.6   Unwrapping, assembling, mounting and setup of the machine/device
   1.7   Connection of the machine/device to supply and disposal networks (water, electricity, computer network etc.) and operation test
2 Operation and usage of the machine/device
   2.1   Initiation of the machine/device
   2.2   Functions of the machine/device during normal operation, safety notes and warnings
   2.3   Refilling consumptive materials
   2.4   Cleaning the machine/device
   2.5   Preventive maintenance (maintenance, inspections)
   2.6   Disposal of supporting and operating materials
   2.7   Shutting-down the machine/device

3   After operating the machine/device
     3.1   Finding the cause of disorder and resolving it
     3.2   Ordering spare parts, wear and tear parts and electric plans
     3.3   Disassembling the machine/device
     3.4   Disposal and recycling of the machine/device (what? where? how?)
4   Appendices
     4.1   Possible causes of disorder/Trouble shooting (what shall I do, if ...?)
     4.2   Spare parts, additional parts (exceeding the supplied/delivered scope)
     4.3   Glossary
     4.4   Index

The above structure pattern for manuals and instructions for use differs from the other structure patterns, because the individual document part headings are partially not as detailed as in the other structure patterns. This was done intentionally, because the described technical products can have very different levels of complexity and very different philosophies of use. Therefore look at this last structure pattern only as an orientation and adopt it to your described technical product.

Naturally-speaking all structure patterns described in this section can be adopted to the described project, product, topic or task. If the supervisor has published an own structure pattern, it should be used. On the other hand, if you use the structure patterns presented in this book, you will establish a correct logical sequence of thoughts, topics, work steps etc.

2.5 Project notebook (jotter)

In a guideline how to write Technical Reports by Thomas Hirschberg, a professor at the University of Applied Sciences in Hannover I found the following advice.

> 🗐 You should *structure the contents of your report as early as possible and note all open problems, decisions and remaining work steps regarding your project "online" in a project notebook (jotter). Do **not** start with writing your Technical Report after all practical work has been completed.*

Since you need your project notebook both at your workplace/in the laboratory and at home, you should use a small booklet in DIN A5 or DIN A6 format.

2.6 The style guide advances consistency in wording and design

A style guide is **only required for the creation of larger written documents**. It is similar with the documentation manual of a documentation or translation service provider. It has the purpose that within a larger document the same things or ideas have the same names (**terminology**) or that they are displayed in the same way (**layout**), so that the document (or Technical Report) is **consistent** in itself.

Therefore in the Style Guide you can collect your preferred **spelling of words, wording of phrases, special terms** as well as **layout rules**. Let us refer to the first and fourth line of the current paragraph. You have probably noticed that the spelling of the special term style-guide differs from the spelling in the previous paragraph and in the heading. However, this should not happen within the same report. Therefore such rules are listed in a style guide and checked during proof-reading and final check with the function "Find" or "Find and replace" of your word processor. Violations of the rules in the style guide are then corrected. In this way the style guide helps to keep consistency in formulation and design within the Technical Report.

In the following **Checklist 2-7** you can see examples of what can be listed or standardized in a style guide. The checklist shows part of the checklist for this book (the German version). The spelling and layout rules have been defined by the editors and the authors of this book, here the editors have mainly defined the margins and the formatting patterns (corporate design). These formatting rules result in a packed layout. For normal Technical Reports the font size of the standard text should be 11 or 12 pt and the gap between paragraphs 6 pt. All other formats listed in the checklist must be adjusted accordingly. Similar layout rules as we have got from the editors can be found in most institutes and companies.

The **usage of a style guide can save a lot of time and effort**. For example, you can store terms, preferred spellings, standard figure titles, drawings, illustrations, logos, copyright notes etc., which appear several times in your Technical Report, in your style guide and copy them where needed from your style guide to your current text file. It is strongly recommended, that you create and use an own style guide for your Technical Reports. It is too time-consuming and insecure, to try to keep in mind all defined rules and regulations.

Checklist 2-7   Example entries in a Style Guide for a Technical Report

Bullet lists and ordered lists always start at the left document margin.
- the first list mark is a small thick bullet (Alt + 0149 while Num Lock key is pressed)
  - the second list mark is a dash (Ctrl + Minus in the numeral keyboard)

Behaviour guidelines (you-form, the hand symbol is not italic, indentation 0,5 cm):
🖰 *Write down ...*

Spelling

use	_do not use_
Words and phrases	
design report	design-report
PowerPoint	Powerpoint

Self-defined formatting patterns

Element	Formatting pattern	Keywords
above list	ListAbove	2 pt empty line
below list	ListBelow	6 pt empty line
figure heading	HeadingFigure	9 pt/ distance above 24 pt/below 12 pt
formula	Formula	10 pt, indentation 1 cm
handwriting	Handwriting	Segoe Script 9 pt/bold/indentation 0.5 cm
standard paragraphs	Standard	10 pt/distance below 4 pt
table heading	HeadingTable	9 pt/distance above 12 pt/below 12 pt Double line ¾ pt
document part	Heading 1	16 pt/bold/distance below 30 pt
headings	Heading 2	13 pt/bold/distance above 18 pt/below 9 pt
level 1 to 3	Heading 3	11 pt/bold/distance above 12 pt/below 4 pt

Entering symbols via numeral keyboard (switched on with FN+NumLock or FN+F11)

Standard font:		– Alt-0150	Symbol font:		⇒ Alt-0222
© Alt-Ctrl-C, Alt-0169	•	Alt-0149	• Alt-0183	×	Alt-0180
® Alt-Ctrl-R, Alt-0174	·	Alt-0183	≈ Alt-0187	≤	Alt-0163
~ Alt-0126	«	Alt-0171	≅ Alt-0064	≥	Alt-0179
± Alt-0177	»	Alt-0187	≠ Alt-0185	™	Alt-0212

soft line break: Ctrl-Minus   dash: Ctrl-Minus in the numeral keyboard

This completes the planning of your Technical Report. Now the most extensive part of the work steps on your Technical Report will be described, i. e. the practical realization of your plans. This contains literature research and reading, writing text, creating figures and tables as well as the continuous adoption of your structure to the current state of your project and development of your Technical Report.

3 Writing and creating the Technical Report

In this chapter you will get many tips and see many examples for the appropriate creation of the Technical Report. Hints for working with word processor systems are mainly collected in sections 3.7.1, 3.7.4 and 3.7.5. However, before showing the details of chapter 3, we want to present some general and summarizing thoughts.

We have already discussed that creating the structure of the Technical Report is the difficult and creative part of the whole task. The structure determines, whether the Technical Report has a comprehensible inner logic.

Despite the fact, that many beginners in "Technical Writing" find it difficult, creating the complete Technical Report is more or less a craft. It includes keeping the rules which are introduced and explained in detail in this book.

Often you can see it in Technical Reports from when on the time-pressure in the project has raised, i. e. from when on errors and un-precise descriptions show up more frequently! Therefore the final check is a very important work step, which may not be left out due to time-pressure at all, see section 3.9.3.

It may happen that a supervisor or customer does not have contents-related reasons to criticize your work. If he/she still wants to find something, he/she often criticizes little odd details or formal aspects. To avoid these problems you should apply the tools report checklist and style guide – both tools have proven to be useful in many projects in practice.

Often institutes, companies, authorities, and other institutions have rules for the optical appearance and computer-based creation of letters, reports, overhead slides, and other documents, so that they fit into the unique optical appearance (corporate design) of the institution. Such rules and guidelines should be integrated into the style guide (section 2.6) and the report checklist (section 3.9.1) and applied during the whole process of creating the Technical Report.

After finishing the phase planning the Technical Report we will now go into the details of writing and creating the Technical Report. In the network plan this phase is marked in gray again.
Please keep in mind the following rule for all tasks marked in the network plan:

> *From time to time you should imagine to be the reader and ask yourself: When does the reader need which information? Does the current figure appear "out of the blue"? Should I pick up the structure, write an intermediate summary, or announce the new document part from a very general point of view? Is the subdivision of information logical and comprehensible?*

L. Hering, H. Hering, *How to Write Technical Reports*,
DOI 10.1007/978-3-540-69929-3_3, © Springer-Verlag Berlin Heidelberg 2010

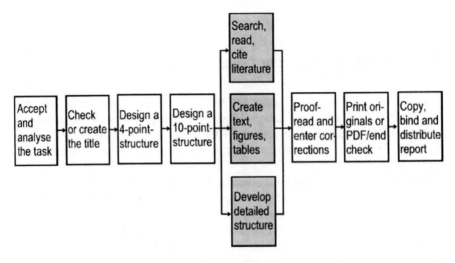

Prior to describing the single steps to create the report in detail, we will provide you with an overview of the general structure of a Technical Report with all parts that need to be written.

3.1 Parts of the Technical Report and their layout

The names, contents and order of the parts of a Technical Report in general are defined in ISO 7144 "Documentation – Presentation of theses and similar documents", **Checklist 3-1**. In Germany this is standardized in DIN 1422. The order of the required parts of the Technical Report as defined in DIN 1422 is slightly different from the order defined in ISO 7144.

Checklist 3-1   Parts of a Technical Report or a thesis according to ISO 7144

front matter
– outside and inside front cover (cover pages 1 and 2)
– title leaf
– errata page(s)
– abstract
– preface
– table of contents
– list of illustrations (figures) and list of tables
– list of abbreviations and symbols
– glossary

body of thesis
– main text with essential figures, illustrations and tables, list of references

annexes
  tables, figures, illustrations, bibliography etc.

end matter
– index(es)
– curriculum vitae of the author
– inside and outside back cover
  (cover pages 3 and 4)
– accompanying material

Not all parts are necessary or required in all Technical Reports. It is the writer's duty to ask the supervisor or customer which rules and guidelines must be kept as long as they are not available in written form.

In the following, we will introduce the individual parts of the Technical Report and give some hints regarding their layout.

3.1.1 Front cover sheet and title leaf

After the "best" title has been developed in section 2.3, the absolute and relative position of all parts that must appear on a front cover sheet and/or title leaf must be defined. A front cover sheet and/or title leaf is a must for a Technical Report.

You should distinguish the (inner) title leaf from the (outer) front cover sheet. The **front cover sheet** is the title visible when the Technical Report lies on a table as a closed book. The **title leaf** is only visible after you opened the Technical Report and in most cases after you turned a blank white sheet of paper.

However, if the Technical Report is bound so that the **(outer) front cover sheet is a transparent sheet of plastics**, then inner and outer **title are identical**, i. e. the information provided on the (inner) title leaf and the (outer) front cover sheet are identical. Then the blank white sheet of paper which usually follows the front cover sheet will follow the title leaf instead.

Beside this special case the following rule holds true: the title leaf always contains more information than the front cover sheet. For instance, in Technical Reports written during study courses it is unusual to list the supervisors on the front cover sheet, whereas they definitely have to be listed on the title leaf.

There are some faults which occur quite frequently on front cover sheets. Some of them are displayed in **Figure 3-1**.

The faults occurring most frequently on front cover sheets are:
• The name of the institution is missing on the top of the page.
• The name of the university is correctly specified, but the name of the department and/or institute are missing.
• The title (essential!) is layouted with a too small font size, while the type of report (not so important!) is much larger than the title.

The layout of the front cover and title leaf are also influenced by more general rules. For example, the corporate design of a company or university may define that for a special type of Technical Report a specific form, e. g. "Cover for laboratory reports" must be used.

In most cases there are also rules for the layout, e. g. that the company logo must always be located at the top and on the right or left side or in the middle. Other layout rules define which font type and font sizes must be used. In universities these rules may exist for an institute, a department, or for the whole university. Naturally speaking, you should follow these rules.

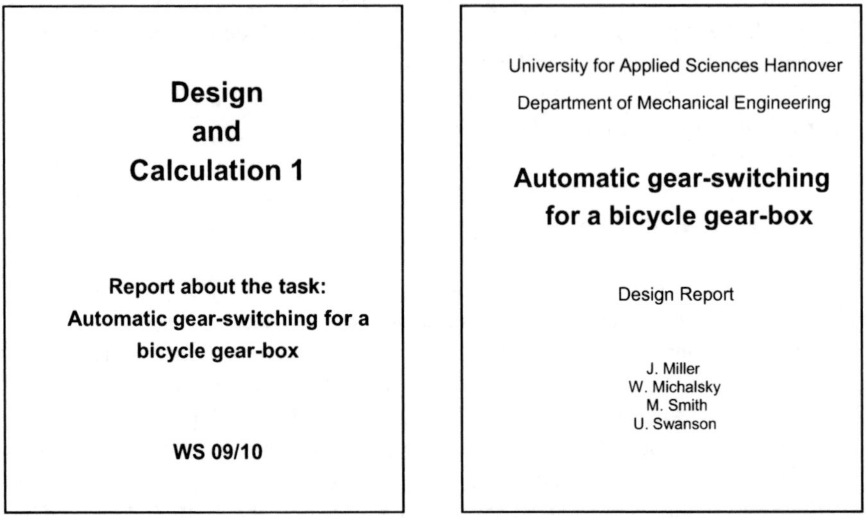

Design **and** **Calculation 1** **Report about the task:** **Automatic gear-switching for a** **bicycle gear-box** **WS 09/10**	University for Applied Sciences Hannover Department of Mechanical Engineering **Automatic gear-switching** **for a bicycle gear-box** Design Report J. Miller W. Michalsky M. Smith U. Swanson

Figure 3-1 Comparison of a faulty (left side) and a correct (right side) front cover sheet
         for a design report

The right version in **Figure 3-1** is fine as long as it does not disobey existing rules of the university or customer. Now we want to look at the steps how to develop the front cover and the title leaf of your Technical Report. The example to explain the work steps is again the dissertation "Computer-aided material selection – CAMS in design education". Our doctorate candidate has got the information from his university which rules should be followed when designing a front cover and title leaf, and he has looked at other dissertations in the university library to see good examples for the application of these rules. The following information must occur on the title leaf of a dissertation:

• the title of the work with the additional specification
• "Dissertation to qualify for a doctorate degree at the University of Klagenfurt",

- the names of both supervisors and the author with full academic titles as well as the city, where the university is located, together with month and year when the dissertation is submitted.

There are no exact rules for the positioning of the information on the title leaf. Therefore the doctorate candidate has manually written down four variants how his title leaf could look like, to judge the line breaking of the single information blocks and to get an impression of the proportions of text and intermediate white space, **Figure 3-2**. The doctorate candidate then types his favourite version into his word processor. There the following typographic design options are optimized:

- Font type and font size,
- methods to emphasize text like bold, italic, expanded spacing etc. and
- justification of the text blocks: centred, left-justified, right-justified, or along a line.

The title leaves for reports and theses written during a study course like diploma thesis, project reports, laboratory, and design reports contain slightly different information from the information in this example. In case of a diploma thesis the student ID number and start and end date of the diploma project must be specified. In case of project reports or other university-internal reports the student ID number and the term or semester are specified. Please refer to the front cover and the title leaf of a diploma thesis, **Figure 3-3**, and a design report, **Figure 3-4**.

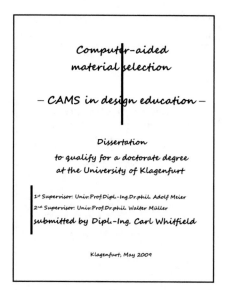 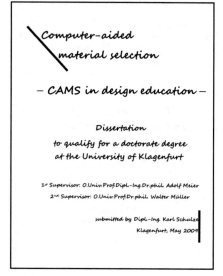

Figure 3-2 Four handwritten drafts of the title leaf of a dissertation (the placement of information varies between centred, left-justified, along a line, and right-justified)

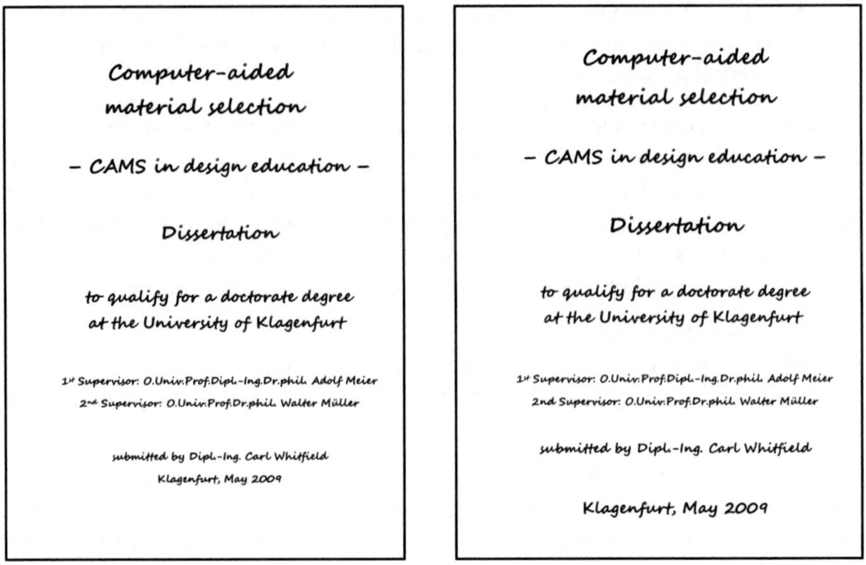

Computer-aided
material selection

– CAMS in design education –

Dissertation

to qualify for a doctorate degree
at the University of Klagenfurt

1st Supervisor: O.Univ.Prof.Dipl.-Ing.Dr.phil. Adolf Meier
2nd Supervisor: O.Univ.Prof.Dr.phil. Walter Müller

submitted by Dipl.-Ing. Carl Whitfield
Klagenfurt, May 2009

Computer-aided
material selection

– CAMS in design education –

Dissertation

to qualify for a doctorate degree
at the University of Klagenfurt

1st Supervisor: O.Univ.Prof.Dipl.-Ing.Dr.phil. Adolf Meier
2nd Supervisor: O.Univ.Prof.Dr.phil. Walter Müller

submitted by Dipl.-Ing. Carl Whitfield

Klagenfurt, May 2009

Figure 3-2 Four handwritten drafts of the title leaf of a dissertation (the placement of information varies between centred, left-justified, along a line, and right-justified) <continued>

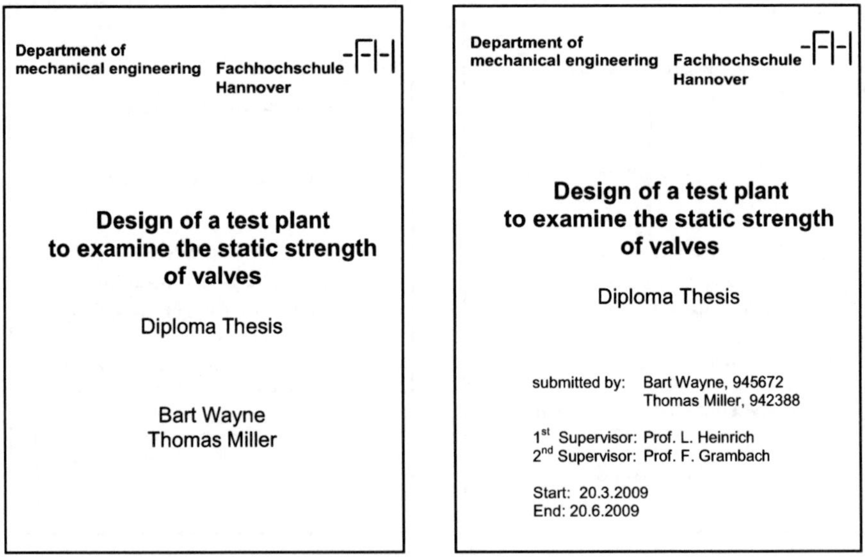

Department of
mechanical engineering

Fachhochschule
Hannover

FH

**Design of a test plant
to examine the static strength
of valves**

Diploma Thesis

Bart Wayne
Thomas Miller

Department of
mechanical engineering

Fachhochschule
Hannover

FH

**Design of a test plant
to examine the static strength
of valves**

Diploma Thesis

submitted by: Bart Wayne, 945672
Thomas Miller, 942388

1st Supervisor: Prof. L. Heinrich
2nd Supervisor: Prof. F. Grambach

Start: 20.3.2009
End: 20.6.2009

Figure 3-3 Front cover sheet and title leaf of a diploma thesis

Figure 3-4 Front cover sheet and title leaf of a design report

The following **Checklist 3-2** summarizes again the minimum information that must be provided ("What") and their location on the front cover sheet and title leaf together with a qualitative specification of the font size ("How").

Checklist 3-2   Minimum information on front cover sheet and title leaf

<u>**Front cover sheet**</u> **for all types of Technical Reports:**

(Logo and) institution
Title of the work (large!)
Subtitle (if applicable)
Type of report (smaller!)
Author/s (medium)
Characteristic image or illustration (if applicable)

<u>**Title leaf**</u> **for all Technical Reports in a study course beside final theses**

(Logo and) institution: university/department/institute
Title of the work (large!)
Subtitle (if applicable)
Type of report (smaller!)
in the subject <name of the subject>
Specification of semester or term (e. g. SS 09)

supervised by: written by: name/s or group and group number
(name with title/s)   (first name/s, name/s, student ID number/s)

Title leaves for final theses

(Logo and) institution: university/department/institute
Title of the work (large!)
Subtitle (if applicable)
Type of report (smaller!)

1st Supervisor: written by:
2nd Supervisor: (first name/s, name/s, student ID number/s)

Start: (exact date)
End: (exact date)

Title leaves for Technical Reports in industry

(Logo and) company, main department, department
Title of the work (large!)
Type of report (smaller!)

written by

Author/s (title/s, first name/s, name/s, department/s, evtl. e-mail, telephone, fax, evtl. addresses of contact persons, promoters, sponsors etc.)

Date and evtl. version (e. g. June 2009 or Version 1, June 2009)

This completes the description of which information is placed where in which layout on the front cover sheet and the title leaf. You have seen different designs in the examples above. The following **Checklist 3-3** summarizes again the work steps to design the front cover sheet and title leaf from sections 2.3 and 3.1.1.

Checklist 3-3   Placement of information on front cover sheet and title leaf

Work steps to place the information on a front cover sheet and title leaf:
- create several variants, use handwriting on paper to avoid restricting your creativity by a limited screen
- try out different line breaks
- form different blocks of information (title, supervisors, company/university, date)
- arrange these blocks centred, left-justified, right-justified or along an angular line
- select the "best" arrangement
- transfer it to your word processor and optimize it there
- care for layout rules of your university, institute, or company

Behind the front cover sheet and title leaf the task, declaration in lieu of an oath (for bachelor, master and diploma theses etc.), acknowledgements, and preface may occur. The next part of the Technical Report is always the table of contents.

More details about what information has to be placed on front cover sheet and title leaf can be found in ISO 1086 "Information and Documentation – Title leaves of books" and in ISO 7144 "Documentation – Presentation of theses and similar documents" and in ISO 5966 "Documentation – Presentation of scientific and technical reports". ISO 5966 contains a visual example of how to arrange the information of a numbered scientific report on the page.

3.1.2 Structure with page numbers = Table of Contents (ToC)

In subchapter 2.4 "The structure as the 'backbone' of the Technical Report" it has been stated, that the document part headings in the structure contain the inner logic of the Technical Report. The structure defines the sequence and the logical super- and subordination of the document part headings. However, in that shape it is not yet suited to look up and search specific passages in the text. Only after adding page numbers the structure becomes a table of contents and then it is suited to look up text passages. Therefore the table of contents of a Technical Report must *always* have page numbers for all document part headings from level 1 to 3 (4).

In this section we will deal with the layout and formal design of the table of contents on the paper. By the way, the table of contents of this book is a good "guideline" for *your* tables of contents.

The headline of the table of contents is *not* – as frequently written – "table of contents". The headline is just "Contents". The fact that it is a "table of ..." becomes clear at first sight on any page of the table of contents. Now we want to present some thoughts regarding page numbers and page numbering.

The **page numbering always begins on the** *first text page*. This is the page, which displays the chapter number "1". In seldom cases the chapter number "0" may occur, if this chapter is a foreword, a preface, an introduction, or other division of similar type, see ISO 7144. In front of the first text page there may occur other pages according to **Checklist 3-1,** especially the table of contents. These parts of the Technical Report are called *front matter*.

Whether the parts of the front matter occur in the table of contents at all and whether they occur with or without page numbers is treated very different. Nearly every book uses different rules for this problem. Therefore we want to make a proposal, how you can solve this issue in your Technical Reports.

• The front matter can get Roman page numbers or no page numbers at all. If the front matter gets page numbers, the title leaf is the first page of the front matter. It is integrated into the page numbering with Roman page numbers, but it does not get a page number printed onto the page. If you apply **book page numbering** with page numbers on the front and reverse side of the pages, the reverse side of the title leaf also does not get a page number. Therefore the printed page numbers start with III on the first page of the foreword/preface or table of contents.

- If you apply the common **report page numbering** with page numbers only on the front side of the pages, the first page of the foreword/preface or of the table of contents, which follows the title leaf, will get the page number II. The rest of the table of contents and the other parts for the front matter will also get Roman page numbers. However, in small and medium-sized Technical Reports the front matter should not get page numbering.

- The **table of contents (ToC)** shall list the parts of the **front matter**, but without page numbers, so that in the table of contents there are only Arabic page numbers. The Roman page numbers of the front matter are much wider than the Arabic page numbers of the normal text chapters. The table of contents should also list the parts of the Technical Report which occur behind the indexes and appendices, like "Curriculum vitae" and "Declaration in lieu of an oath" in dissertations in the right order, but without page numbers.

- ISO 7144 establishes the following rules regarding page numbering:
  - Title leaves are integrated into the page numbering, but they do not get a page number.
  - The pages are consecutively numbered with Arabic numbers, empty pages and the front matter are counted as well. The first page number occurs on the front side of the first printed page.
  - The pages of the annexes/appendices will get own page numbers with Arabic numbers which contain the letter of the annex/appendix and the page number starting from 1.

In the table of contents the page number being listed is **only the first page number** of any document part. A frequently occurring mistake in Technical Reports is to list start and end page number of the document parts with an extension mark in between.

Thus, the following ToC entry is wrong:
5.1 Experiment set-up 35-36

Correct is:
5.1 Experiment set-up 35

The page numbers in the table of contents are printed right-justified.

After the placement of the page numbers is defined, the placement of the document part headings still needs to be discussed. ISO 2145 "Documentation – Numbering of divisions and subdivisions in written documents" provides a layout example for a table of contents where independent of the hierarchy level in the document all document part numbers are aligned along a common building line. All document part headings are aligned along another common building line more to the right. **Indentations are not recommended in ISO 2145**. This kind of layout is shown in the following **Figure 3-5**.

5	Welding experiments with the optimized extraction system	34
5.1	Experiment set-up	35
5.2	Preparations for the experiments	36
5.2.1	Definition of the experiment program	38
5.2.2	Used equipment	41
5.2.3	Used materials and expendables	45
5.3	Experiment execution	47
5.3.1	Program flow chart for the experiment execution	50
5.3.2	Additional remarks regarding the experiment execution	51
5.4	Experiment evaluation	52
5.4.1	Experiment evaluation based on the weld seam quality	53
5.4.2	Experiment evaluation based on the suction system effectivity	57
5.4.3	Assessment of handling and visibility conditions	59
5.4.4	Evaluation of flaws	60

Figure 3-5 Table of contents of a chapter according to ISO 2145

However, since many decades tables of contents are often layouted with indentations, as it is shown in the following **Figure 3-6**.

5	Welding experiments with the optimized extraction system			34
	5.1	Experiment set-up		35
	5.2	Preparations for the experiments		36
		5.2.1	Experiment program and used equipment	38
		5.2.2	Used materials and expendables	45
	5.3	Experiment execution		47
		5.3.1	Program flow chart for the experiment execution	50
		5.3.2	Additional remarks regarding the experiment execution	51
	5.4	Experiment evaluation		52
		5.4.1	Experiment evaluation based on the weld seam quality	53
		5.4.2	Experiment evaluation based on the extraction system effectivity	57
		5.4.3	Assessment of handling and visibility conditions	59
		5.4.4	Evaluation of flaws	60

Figure 3-6 Table of contents of a chapter with indentations for a better overview

A structure or a table of contents with indentations is much clearer and is therefore recommended!

To achieve this result you should use indentations and tabs. If you use space characters, it is not possible to keep the vertical building lines precisely. If each level in the document hierarchy starts at an own building line, the reader can comprehend the inner structure of the Technical Report much better. And the author can constantly check the logic of the report when writing the 4- and 10-point structure and the detailed structure.

Next it will be shown how much the checking of the logical order of document part headings is facilitated by the indentations. Look at **Figure 3-6** and read the document part headings on level 2 along their building line. You can read the following terms: "Experiment set-up", "Preparations for the experiments", "Experiment execution" and "Experiment evaluation".

A check of the inner logic ("backbone") results in the following thoughts. After the description of the experimental equipment there is a description of the preparations which have to be executed before the experiments can be started. Then there is a description of how the experiments are executed and an evaluation of the measured results. Conclusion: the inner logic is properly built up!

These constant checks for the logic of the report during writing are effectively supported by the indentations in the structure and the table of contents.

In ISO 2145 all document part headings are arranged along a common building line. This makes it harder to optically recognize the document part headings of the same hierarchy level which belong together. When looking for the next document part heading of the same hierarchy level you have to read the next headings one after the other, check their hierarchy level, and eventually reject them. If e. g. the headings of the second hierarchy level shall be checked, and along the common building line there are headings of the first and third hierarchy level as well, the reader has to execute complicated comparing and sorting processes in his brain, which require an unnecessary amount of memory capacity. These processes are superfluous and can be easily avoided by a better alignment of document part headings. Beside indentations the structure of the document part headings can also be optically emphasized by other means. Chapter headings are often printed in bold type in the table of contents, while the document part headings of lower hierarchy levels aren't. The hierarchy level can also be expressed by the font size, e. g.

- chapters 14 pt,
- subchapters 12 pt and
- sections 11 pt.

Document part headings should **never be printed in upper-case letters only** (capital letters or small capital letters), because the eye is not used to it. Thus, the headings are much harder to read. The reason is, that during reading eye and brain process the word contours like a picture as a "skyline" and compare them with formerly stored "skylines". If capital letters are used, the letters have to be read one after the other and the meanings of the words have to be analyzed.

Moreover **groups of document part headings** can be built up by a **variation of the vertical distance** between the document part headings in the table of contents. Leading characters (dots) should be used between document part heading and appertaining page number to facilitate reading. **Figure 3-7** shows an example, how these mechanisms can interact effectively.

Contents

Figure 3-7 More typographic accentuations make the structure of the table of contents even clearer

If a document part heading does not fit on one line any more (because it is too long or indented), it must be continued in the next line/s at the appropriate building line for the text of the document part headings of the relevant hierarchy level. The leading characters (dots) and the page number appear in the *last* line of this document part heading. The page numbers should be formatted with the same font size without any accentuations, **Figure 3-7**.

☞ It is clearer and more pretty, if you insert a space *between the document part heading in the table of contents and the tab with the leading dots as well as between the tab and the page number. In Word, you can search for the tabs with Edit – Search – Extended – Other or search for "^t" and replace the tabs with space, tab, space.*

The **gap between all document part headings and the** appertaining **page number** should be **filled with the same leading dots**. The tab with the leading dots should not be formatted in bold face typing. This is the same **for chapter headings**. The font size should also be the same in the gaps, because dots have a larger distance and diameter at 14 pt font size compared with 10 pt, **Figure 3-8**.

...

   3.2 Command sequence for the installation 5

4   Program usage 6

   4.1 Submenu "File" 6

      4.1.1 File – New 7
      4.1.2 File – Open 7

...

Figure 3-8 Leading dots (and page numbers) in bold type and different font size should be avoided

Subheadings without an own document part number do never occur in the table of contents. Examples are subheadings marked with "a), b), c)", "α), β), γ)", or just with bold type. Unnumbered subheadings are used, e. g. to avoid a subdivision into the fourth level. They can also occur within calculations.

Different from document part headings, subheadings may have a **colon at the end**, if they have an **announcing character**.

If there are four or more hierarchy levels, the indentations reach relatively far to the right. Therefore many document part headings can stretch across more than one line. In addition the vertical distance to the superordinated document part heading becomes quite large. To avoid these problems there are several options:

- The table of contents may be layouted in a smaller font size than the normal text, e. g. normal text in 12 pt and the table of contents in (10 or) 11 pt.
- Subheadings without document part numbers can be used (see above).
- Subheadings with document part numbers are used, which do not occur in the table of contents (actually this should not happen!) in Technical Reports, see next item.

- Subheadings with document part numbers are used in the text, which do not occur in the overall table of contents. However, they are listed in a detailed capitular table of contents. This approach is sometimes applied in larger documents (e. g. in manuals and textbooks), but it is rather unusual for Technical Reports.
- The indentations are smaller than the widest document part number. This causes less document part headings to stretch across more than one line.

☞ *If such problems in the table of contents occur, the author should discuss with his supervisor or customer, which variant is to be used. In doubt the classical solution with an overall table of contents in the front of the Technical Report showing all numbered document part headings should be preferred.*

Since the options to format an automatically created table of contents after its creation are partly not so well-known, in practice the **tables of contents** are either created with the default layout settings, which looks ugly and is not very clearly arranged, or **typed as normal text** and layouted accordingly. If they are manually typed, **mistakes can creep in**. You should especially look at these points:

- All document part numbers and document part headings in the table of contents must be exactly consistent with those in the text. Each document part heading gets its own (begin) page number in the table of contents. Violations of this simple rule can be found in many reports!
- During the end check prior to the final printout you should check, whether all numbered document part headings are listed in the table of contents with the right document part number, the right wording, and the right page number! If you let the computer create the table of contents, you should update it once again now, see 3.7.4 "Automatic creation of indexes, tables, lists, labels and cross-references with Word".

3.1.3 Text with figures, tables, and literature citations

The "text" contains all information presented in the chapters (e. g. starting with *Introduction* and ending with *Summary*). In the text you will also find tables and figures, formulas and literature citations. Information which is not written by the author but cited as a base for the author's ideas or argumentations, must be clearly marked as a citation, see 3.5. This is also necessary for cited figures and tables.

The individual parts of the Technical Report like tables, figures, literature citations, text, and formulas are described in more detail in subchapters 3.3 to 3.7. Here in the beginning of the chapter some aspects shall be discussed, that cannot be assigned to the more detailed subchapters because of their general relevance.

The **author** of a Technical Report **shall guide the reader with his words**. All intermediate thoughts, conclusions, etc. shall be explicitly communicated to the reader in the text. **Thus, the reader can follow all logical thoughts of the author** regarding the sequence of the report **and follow the thread in his own mind**. This improves the understandability of the Technical Report very much.

However, if a figure, a table or a bullet list follows directly upon a document part heading, the reader is often somewhat left alone. In most cases an "introductory sentence" is missing here, which announces and explains the logic thread or the contents of the figure, table, or bullet list.

This deficiency occurs so often, that the authors use the abbreviation "*is*>" at the left margin. The peak of the angle points to the gap between document part heading and figure, table, or bullet list and the abbreviation denotes: the "introductory sentence" is missing here. Look at the following example:

… <Text> …

3.3 Morphological box

The sub functions identified in 3.2 "Break-up into sub functions" and the mentally designed solutions for the sub functions are now clearly arranged in the morphological box..

Sub functions	Solutions for the sub functions			
	1	2	3	4
A Create rotation	Electric motor	Diesel engine		

… <Text> …

This introduction guides the reader. So he cannot lose the overview across the sequence, which improves the understandability of the Technical Report.

The chapters "Introduction" and "Summary" are of major importance for the Technical Report. These two chapters are the first ones most readers scan after a quick look into the title and table of contents, before they start to read the text thoroughly. These chapters are introduced in **Checklist 3-4** together with examples of structure and contents.

Checklist 3-4   Introduction and Summary

The introduction
- is located at the beginning of the text and is normally the first chapter.
- describes the starting situation at the beginning of the project, the relevance of the project for the particular field of science, the results of the research for the society, and other similar aspects.
- can contain a description of the project task and project target formulated by the author.
- can also contain thoughts related with the project regarding the following topics: economics, technology, laws, environment, organization, social care, politics, or similar topics.
- should tie up with prior knowledge and experiences of the readers.

The summary
- is located at the end of the text and is normally the last chapter.
- can have document part headings like: summary, summary and conclusions, summary and evaluation etc.
- discusses the task. Hence: What should be done and what has actually been reached, where have been special difficulties and which parts of the task could eventually not be treated and why.
- normally describes shortly what is covered in which chapter and subchapter of the Technical Report (picking up the structure!). When writing these sentences, you should express how the document parts are logically related with each other (starting with ..., then ..., next ..., due to ...).
- can give advice for a reasonable continuation of the project or scientific work in the conclusion. Such advice is normally based upon experiences made during the work on the current project.

3.1.4 List of references

The (overall) list of references is normally put directly after the last text chapter and lists all references to the literature from which there are citations in the whole Technical Report. In larger documents (e. g. in manuals or textbooks) there may be a capitular list of references after each chapter. The appendices follow the overall list of references or the capitular list of references of the last text chapter. It is unusual to integrate the list of references into the appendix. The list of references of a Technical Report has its own chapter number and stands alone between text and appendix or appendices.

To present all information regarding working with literature citations together, we have concentrated the why and how to make literature citations and how to design the list of references in subchapter 3.5.

3.1.5 Other required or useful parts

The position of other required or useful parts within the Technical Report and their layout is even more dependent on university or company internal regulations than this is the case e. g. for the list of references.

A bound thesis like a diploma, bachelor or master thesis often contains the **task**. It is written by the institute or the supervisor and bound with the rest of the thesis. It is usually the first sheet after the inside front cover.

In addition, a diploma, bachelor or master thesis requires that the student presents a **declaration in lieu of an oath**. This declaration confirms that the student has written the thesis himself and that all used literature sources, rigs, machines and tools are listed completely and truthfully. The exact wording and the position of the declaration in lieu of an oath within the thesis are generally defined by the university. Beside to diploma, bachelor and master theses, doctorate theses and other final theses also contain such a declaration in lieu of an oath. The declaration in lieu of an oath must be personally signed by the candidate for the bachelor, master, diploma or doctorate degree. In most cases it is even required that the signature may not be copied. In bachelor, master and diploma theses the declaration in lieu of an oath is mostly part of the front matter and follows directly after the task, in doctorate theses it is mostly part of the back matter and is located directly before the Curriculum Vitae (CV).

In bachelor, master, and diploma theses there is often a page with **Acknowledgements**. This is mainly the case, if the project, which the thesis describes, has taken part outside of the university and if there shall be expressed a special thank you to staff members of industrial companies. However, you should not forget the supervisors from the university here. Without their willingness to supervise the project and the writing of the thesis, and without their experiences, which topic is suited in which detail as a bachelor, master, diploma, or doctorate project, many projects would not take part at all or would last much longer than planned. If nothing else is defined or required, the acknowledgements should be put before the table of contents or the preface/foreword, if there is one.

Eventually an **abstract** is required by the institute or company. In any case it may not be longer than one page, better is only half a page. Its heading is "Abstract" (or "Short summary") to distinguish it from the normal "Summary" at the end of the Technical Report. The abstract is put directly in front of the introduction. For most articles in scientific journals an abstract is obligatory before the article text following the title and the names of the authors. This abstract of an article is often in italic type and occurs in English and German or another language.

Books often have a **foreword or preface**, pointing out the information target, changes since the last edition or specific rules how to use the book. A foreword or preface is always located directly before the table of contents.

The layout of the table of contents is described in 3.1.2.

A **List of figures** contains figure numbers, figure titles, and page numbers. A **List of tables** contains table numbers, table titles, and also page numbers. The page numbers should be placed right-justified along a common building line. The distance between the end of the figure or table title and page number should be filled with leading dots as in the table of contents.

The **List of abbreviations** contains the used abbreviations – in alphabetical order – and an explanation for each of them. The explanations should start at a common building line on the right side of the abbreviations. If the explanations have the character of definitions, they can be introduced by equal signs. If the abbreviations are a combination of the first letters of the words in the explanation, these first letters can be emphasized with bold type and evtl. capital letters and underlining.

If a Technical Report contains many mathematical formulas, a **List of used formulas and units** may be helpful. The explanations of the formula symbols should again start at a common building line on the right side of the formula symbols. Depending on the type of your report you may as well create other lists, e. g. for checklists, exercises, link lists etc., see Appendix A of this book.

As a comparison and to summarize it, we want to remind you of the structure of a Technical Report according to ISO 7144:

- **Front matter:** Front cover (cover pages 1 and 2), title leaf, evtl. errata page(s), task (What should be done?), evtl. declaration in lieu of an oath (How has it been done? "on my own"!), evtl. acknowledgements, abstract (What is the result?), foreword or preface, table of contents, list of figures, list of tables, list of abbreviations and symbols, glossary.
- **Body:** text chapters with required figures, tables, formulas, list of references.
- **Appendix/Appendices:** annex material like figures, tables, bibliography, list of standards, evtl. accompanying material, evtl. glossary, index(es).
- **End matter:** curriculum vitae of the author, inside and outside back cover (cover pages 3 and 4), evtl. accompanying material.

Accompanying material like the bill of materials, technical drawings, measuring protocols, slides, models, leaflets, brochures etc. can be presented as independent appendix chapters or as subchapters in an appendix. They can be bound in the Technical Report or delivered in an external container like a second volume, a drawing roll, a folder or a box.

To place figures, tables and measuring protocols into the appendix/appendices is only useful, if placing them in the body would disturb smooth reading of the text chapters too much.

The appendix/appendices in the list above can be independent chapters. Then they get consecutively numbered appendix chapter numbers. If the number of appendix chapters becomes too large, the annex materials, lists, and accompanying materials can be combined in one common **chapter "Appendix"**, see Appendix A in this book. The **Glossary** – if in opposition to ISO 7144 it is put at the end of the Technical Report – **and Index** remain **separate chapters** in any case, see appendices B and C in this book. The page numbers for the Glossary and Index may be just Arabic numbers as in this book or a combination of a letter for the chapter and a number (e. g. page numbers C-1 to C-10) or a combination of the chapter name and a number (e. g. page numbers Index-1 to Index-10).

The following two examples demonstrate that an appendix can be organized as one chapter with subchapters or as several chapters.

Appendix as several chapters	Appendix as one chapter

1   Introduction 3
(2 to 7 = other chapters)
8   Summary and Conclusions 65
9   References 67
A   Abbreviations A-1
B   Figures B-1
C   Tables C-1
D   Important Standards D-1
E   Bibliography E-1
Index Index-1
	1 Introduction 3
	(2 to 7 = other chapters)
	8 Summary and Conclusions ... 65
	9 References 67
	A Appendix A-1
	  A.1 Abbreviations A-3
	  A.2 Figures A-6
	  A.3 Tables A-47
	  A.4 Important Standards . A-65
	  A.5 Bibliography.............. A-67
	Index Index-1

The variant "Appendix as one chapter" has the advantage that the document part numbers remain smaller and clearer. Besides the text chapters are normally subdivided, so that the chapter heading is the superordinated concept. It is more logic in itself to apply this approach in the appendix as well. Therefore the structure on the right is recommended. Naturally speaking, all document part headings in the table of contents get page numbers.

Now we want to give you hints for the design of the different parts of the appendix.

Figures or tables as annex material are often arranged in the appendix due to comfort reasons, because figures and tables can be handled more easily in an appendix than in a text chapter. Such an appendix forces the readers to turn the pages back and forth in the Technical Report very much. The text-figure-relationship of this solution is quite bad. Therefore figures and tables should rather be integrated into the text chapters in the front. However, *design* drawings should be placed in an appendix. If they would all be integrated in the text chapters, this would disturb smooth reading too much. If individual drawings are explained in the text in detail (e. g. Modification of an experiment rig), these drawings may additionally occur in the appertaining text chapter, evtl. as a reduced DIN A4 or A3 copy.

If you *cite* **other materials** (brochures by associations or companies, catalogues by manufacturers, important standards and other literature) *as references*, they *must appear in the list of references*. If they are not cited, it is sufficient to just add them to your appendix as additional information (as original or copy).

Please plan this early and do not write notes into your original materials. You may mark selected dimensions with yellow highlighter.

The pages of the appendices may get consecutive Arabic numbers (continued from the last page of the list of references). However, according to ISO 7144 the appendices get consecutive capital letters instead of chapter numbers and the page numbers consist of chapter letter and a consecutive Arabic number.

Enclosed documents or copies usually have their own page numbers. These existing page numbers remain untouched and every brochure, catalogue, or technical document gets a consecutive document number ("1", "2", "3", … or "Document 1", "Document 2", "Document 3" …), which is glued onto the document with **white adhesive label**. Sticky-notes ("Post-it") are not suited for this purpose.

Usually the enclosed documents are original brochures or photocopies which are bound with the other pages of the Technical Report. If the enclosed documents cannot be bound, because they are too thick, too many or larger than DIN A4, they should be enclosed in a *separate folder or box or roll*. Such *documents or drawings which are delivered separate* from the Technical Report are *listed in the table of contents* of the Technical Report *at the logically right position* and get a comment like "(in drawing roll)" or "(in separate folder)". Example:

If an appendix has a substructure, a **title leaf for an appendix chapter** is very useful. Such a title leaf gives an overview of the appendix and forms a capitular table of contents. For example, Appendix A contains a bill of materials and (bound) design drawings. In the front in the **overall table of contents** this structure would be displayed as follows:

In the back in the **Appendix A** the reader would be reminded of the structure of the appendix by means of a **title leaf**. This creates clarity and overview for the reader.

The layout of the title leaf of an appendix is partly equivalent to the layout of tables of contents (bold type of the chapter heading, leading dots, page numbers) and partly equivalent to the layout of title leafs (generous spread and pleasing arrangement of the printing ink on the paper).

If the appendix/appendices contain many plots, measuring protocols, program listings, drawings and other documents printed on DIN A4 paper, these **annex materials can be bound separately as volume 2**. The table of contents of both volumes should list all contents in both volumes, i. e. volume 1 and volume 2 have an overall table of contents with the sections "Contents – Volume 1" and "Contents – Volume 2".

Now some remarks regarding special appendices that – if they exist – must always be separate appendices following the other appendices or the list of references. They are listed in consecutive order.

A **Glossary** contains technical terms and explanations of these terms. It is helpful, if the Technical Report deals with a specific field and the readers may not completely know the relevant terminology of this field. If the Technical Report is written in English but published in a country with an official language other than English, you should think of adding the technical terms in the official language of the country after the English term. You can add it in brackets or with a dash. To combine the technical terms in the official language with the technical terms in

English facilitates the exchange of ideas in the scientific community, because much of the literature is written in English and many conferences are held in English. The technical terms in the glossary are accentuated by bold or italic type. They are ordered along a common building line. The explanations start either on the right side of the terms at another common building line or in the next line indented by approx. one or two centimetres. Even more space-saving is the version in this book, Appendix B.

An **Index** contains keywords in alphabetical order. It is only useful for larger Technical Reports. The entries must be formulated *from the readers' point of view*. A structuring into superordinated and subordinated concepts (on max. two levels) results in more clarity and better overview. The page numbers are partly added with commas directly after the keywords or they are displayed right-justified along a common building-line. The gap between index entries and page numbers should then be filled with leading dots. This looks more pleasing, especially, if the index has two or more columns. If an index entry occurs on more than one page in the text and on one page there is more or very important information, this page number can be accentuated by bold type and thus it can be marked as main entry.

Doctorate theses (Ph. D. theses) normally have a **Curriculum Vitae** (CV). It lists roughly the previous educational and professional development of the doctorate candidate. In doctorate theses another part in the appendices is the **Declaration in Lieu of an Oath**. Often it is placed behind the List of References or behind the CV. Contents and structure of the CV depend very much on the university or faculty. Also the placement of CV and Declaration in Lieu of an Oath might be different than proposed here. Therefore the doctorate candidate should ask his doctoral supervisor for an example CV and Declaration in Lieu of an Oath, evtl. with anonymous (i. e. unrecognizable) personal data.

3.2 Collecting and ordering the material

Up to now the following parts of the Technical Report have been created:

- exact title with design of the title leaf and
- structure (detailed up to 10-point-structure or finer).

These parts give orientation for collecting the material. They help to answer the following questions:

- What is needed overall?
- What is already available?
- What is still needed?

Collecting the required material and information must be oriented towards the information target and the target group.

- Which knowledge and experiences do I want to impart how?
- Is there a visualisation (figure, table, formula) for each important statement or at least a bullet list?

In most cases, to collect, order, and write something the reader does not need is unnecessary work!

Now we want to describe how the **material collection** can be done in practice. All spontaneous **thoughts and ideas regarding the contents of the Technical Report** should be collected on one or more sheets of paper regardless of their sequence and their assignment to already existing items in the structure. Evtl. you want to note **each idea separately on note sheets**, writing paper or file cards e. g. in DIN A5 landscape format. Then you should write on the front sides only, so that you can spread the sheets or cards on the desk or floor to sort and order them.

The material collection should include information, in which section of the own Technical Report which internet sources, books and other references shall be referred to or cited. Right from the beginning, please write down all required bibliographical data exactly with page number(s), section number or exact URL (internet address and date when the information was accessed), so that when you start writing you do not need unnecessarily long time for (re-)searching what you had "digged out" before and you can cite correctly.

If a Technical Report is more comprehensive, i. e. it has **more than 20 pages**, the material collection should not be done for the whole report. In this case it is better to execute a separate material collection **for each chapter** or subchapter.

⌂ *Ordering the material can be executed in different ways. In this phase you should again have your target group in mind:*
- *Are all noted items interesting for the readers? (if not: cross them out)*
- *Can the noted items be already assigned to existing items of the structure? (If not: Open a new item in the structure)*
- *If you took your notes one after the other on Din A4 paper, you should mark your ideas according to their "sequence". You can, for example, use the document part numbers from the structure and consecutive numbers within one section. If you noted each of your ideas on separate sheets or cards, you can order the sheets or cards according to their "sequence".*

The "sequence" of thoughts can follow a chronological or logical order (by starting conditions, targets, alternatives, components, fields of knowledge, branches of science etc.). This chronological or logical order is – at least partly – predefined by the already existing structure of the Technical Report. Therefore, ordering the material is only possible, if a 4-point- and 10-point-structure have been previously created. Due to the order of work steps recommended here, collecting and ordering the material will automatically be logical und oriented towards the information target. This method saves working time, because all work results fit together and only a few of them will end up in the "wastebasket".

3.3 Creating good tables

Tables display information in a matrix of rows and columns. The fields in this matrix are called **cells**. Tables often have a header and an introductory column. The **header** contains the **superordinated concepts (generic terms) of the columns**. The **introductory column** contains the **superordinated concepts (generic terms) of the rows**. The terms in the header and in the introductory column should as much as possible be consistent in themselves and logically equal.

↓ **introductory column**

header →

Product	Unit price	Complete price (4 pieces)
Winter tyres	49,30	189,90
Summer tyres	46,50	179,90
Steel wheels	28,90	109,90
Aluminum wheels	126,30	479,90

The **upper left cell** in a table which belongs to the header and the introductory column can contain the following:

- both superordinated concepts of header and introductory column
- only the superordinated concept of the introductory column
- no entry (empty)

If both superordinated concepts are short, this cell may be separated by a diagonal line from the upper left to the lower right corner. However, if the terms are longer, you should use arrows to indicate which term belongs to the header and the introductory column, **Figure 3-9**. Diagonal line and arrows can be created as graphical elements.

In **smaller tables** the impression overweighs, that **information can be displayed very systematically, well-arranged and structured** in tables. However, **larger tables** are often **confusing** due to their poor amount of visualisation. If the table displays words, this is less problematic than if it displays figures.

Load Type	10 N	20 N	30 N
A	2 mm	4 mm	6 mm
B	1 mm	3 mm	7 mm
C	2 mm	5 mm	9 mm

Voltage ⟶ ↓ Potentiometer ↓ position	50 V	150 V	250 V
Position 1	6 A	18 A	30 A
Position 2	8 A	24 A	40 A
Position 3	12 A	36 A	60 A

Figure 3-9 Different ways to structure the upper left table cell

For readers, who read a **figures-table,** it is often difficult to estimate relations of numbers and to compare sizes. Therefore it is often required to **visualize figures-tables** by means of **diagrams (charts)**.

Yet, you still have to proof calculations, statistic analyses, and experiment results with exact figures in your Technical Report. Therefore you should consider the following compromise: Bulky and confusing figures-tables are provided in an appendix. In the text chapters there are well arranged visualisations (diagrams, charts) and the figure titles also refer to the related table in the appendix, e. g. with the annotation "(see table xx, page yy)".

In integrated office program packages the word processing and spreadsheet programs are compatible, figures (or tables resp.) can be easily exchanged between them. In such programs the figures in the table can be displayed in a selectable diagram type as presentation chart. For example in Word there is such an integrated presentation graphics program available, which allows that you use the same figures in the back in an appendix as a table and in the front in a text chapter as a presentation graphic. Usual and often-used diagram types are line, column, and circle diagram. Please refer to 3.4.5 "Scheme and diagram (chart)" to find tips for the selection of the diagram type and the design of schemes and diagrams.

If a table is large or contains much information, the clarity of the table can be improved by some typographic means which are described in more detail in the next section.

3.3.1 Table design

As already mentioned, figures-tables are especially confusing and nearly every graphic illustration is clearer. Therefore figures-tables should be carefully structured with suited measures.

The German standard **DIN 55301** "Design of statistic tables" defines that the design of tables should be spare, **horizontal and vertical lines shall only be applied for the header and the introductory column**. At the lower end the table shall not get any boundary. Here is an example for table design according to DIN 55301. The displayed data is derived from an article by Dirk Schmaler, "Manche Klimaanlage hat ungeahnten Durst" (Some air-conditions are unexpectedly thirsty), Hannoversche Allgemeine Zeitung, Wednesday, 17 May 2006, No. 114, p. 8. They cite the table with "Source: Plusminus/WDR".

Manufacturer, model, motor, year of manufacture	Fuel consumption l/100 km w/o air-condition	with air-condition	Additional consumption Litres	Percent
Daihatsu YRV 1,3, 64 kW/2001	8,64	11,16	2,52	29,1
Mitsubishi Colt 1,6, 76 kW/2001	10,46	12,92	2,46	23,5
Audi A4 1,8, 92 kW/2001	13,32	16,58	3,28	24,4
...				

In Technical Reports there should be at least **a horizontal line at the end of the**

table like it is done at the end of all checklists within this book. To distinguish the table from the text even better you should use a closed outer frame.

The information **within the table** should be structured as well. Horizontal and vertical lines are well-suited to distinguish the rows and columns. An additional option to structure the table is to use **double lines or thicker lines to separate the header and the introductory column**. See the examples below for an inspiration.

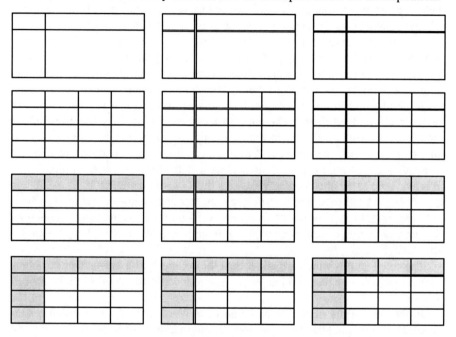

In Word, please have a look into the design options in the menu Table – Auto-Format. In our opinion most of the default settings are not suited for Technical Reports, but you can save your **own table formatting templates** there.

☞ *In any case you should apply one constant table design within one report.*

The lines that form the rows and columns must have a **sufficient line thickness** so that the lines are still visible, even if you copy the table and then take a copy from the copy. In most cases the hairline, as the thinnest line is called in many drawing and word-processing programs, is too thin for table separators. ½ pt is the lowest line thickness you should apply.

Whether a ¼ pt line can be copied without problems is very much dependent on the used equipment (printer and copier). Therefore, if you want to use ¼ pt lines, you should try out printing and reproduction to evaluate the quality of the reproduced copies in advance and adopt the line thickness, if necessary.

The **terms in the header** should be formatted in bold-type in any case. It might be useful to format the **terms in the introductory column** in **bold-type** as well.

You may also apply a larger font size and a colour fill or shading to emphasize the terms in the header and the introductory column.

The **information within the table** can be structured **with blank lines** (additional white space) after max. five text lines.

Other options to emphasize or structure the information in a table are **indentations** of the subordinate information in a column, different font sizes, and **shading** cells. For example, shading is used in the business section of daily newspapers. There the table rows appear alternating on gray and white background.

The shading should not be too dark so that there is always sufficient contrast with the text which is usually printed in black. In the printer driver (File – Print – Properties) you can influence the colour saturation of your printer. There you can also control brightness and contrast and activate toner saving mode.

If the report shall be copied, you have to keep in mind that **shaded areas are sometimes changed during the copy process**. Such areas may become **spotted or blurry** sometimes already in the first copy, contrasts may be emphasized, so that dark areas become darker and bright areas disappear. Therefore you should copy all pages with gray and coloured areas with the halftone or photo function, if it is available at the photocopier. Please try that out *early enough*!

Write down all page numbers that require "special treatment" (photo function, different brightness than normal, zooming etc.) into a checklist for reproduction, so that you do not process these pages with the rest of the stack, but copy them individually!

Table cell entries which are wider than the column width must either be abbreviated to stay in one line or a line break is introduced which results in a larger cell height.

In the other cells of the current line of a Word table the text remains top-justified. If this is not the case, you should reformat the cells to be top-justified. In Excel and the spreadsheet program Microsoft Works the default setting is bottom-justified. You may change the horizontal and **vertical justification** of the table entries in Works with Format – Alignment and in Excel with Format – Cells – Alignment.

Equivalent to figures text tables shall be readable **from the normal reading distance of 30 to 40 cm without any problems**. This requires a sufficient font size especially for indices and exponents. If you copy them, parts of these small letters and/or digits may become unreadable. If you are uncertain, you should create a master copy of the page in question and make a copy from it. If there are problems, you should use a larger font size or a larger magnification scale if the table is copied from a book or article or integrated as a pixel graphic file.

The **entries in the cells** of a table can be figures, words, sketches, or a combination of these elements. For text entries the normal rules of case sensitivity are valid, especially adjectives start with small letters. If you like, you can go to the menu Special – AutoCorrect options, tab AutoCorrect and change the setting "Begin every table cell with a capital letter".

The font type used in the table can be a proportional or a fixed spacing font. Here are some recommendations regarding fonts in tables:

- Filling in tables is sometimes easier, if you use **fixed spacing fonts**, because you do not have to work with tabs and indentations all the time. When using fixed spacing fonts indentations and column alignment can be realised by adding or deleting space characters. However, these fonts need much space!
- Tables with **proportional fonts** are the "state of the art", if you use an up-to-date word processor. They fit better with the normal text font and appear typographically smoother.
- **Sans-serif fonts** like Arial and Helvetica are better suited for figures-tables than serif fonts, because without serifs you can recognize the outline of the digits faster.
- **Large figures** can be identified easier if they are formatted with a fixed spacing font like Courier, Sans Serif, Letter Gothic, Monospaced, Lucida Console etc., if the digits before the decimal point are arranged in groups of three digits. If possible, the decimal groups shall be divided by space characters only, not by commas (e. g. 3 155 698).
- Decimal points must be listed one below the other, even if the figures have a different number of digits after the decimal point.
- Due to safety reasons amounts of money are written without space characters or they get commas to identify the decimal groups.
- If you use proportional fonts, use a right tab for columns of integers and a decimal tab for columns of rational figures.
- **Texts** within cells are mostly **left-justified**. A centered text justification often results in a too unsteady layout, especially if the words or texts listed in one column have a very different length.
- Sometimes tables are so wide, that you want to rotate them to the left so that they can only be read in landscape paper orientation. Then the readers have to rotate the bound Technical Report to read the table. If you succeed to keep **portrait orientation** by using a smaller font size or by exchanging table header and introductory column, this is nearly always better readable and the handling is facilitated, because you must not rotate the document. However, this must not result in a too small font size. We recommend to use 9 pt as the smallest font size for every font type.

Often there is a **legend** below a table, that explains the meaning of abbreviations, symbols, line types, evaluation categories (to be calculated as plus or minus values) etc. To assign these explanations to the table you should use the heading "Legend:". This connects the information under or next to the heading with the table without doubt and distinguishes it from the normal text and the table heading.

Another option to explain the table entries is to use **table footnotes**. The reference in the table is often a superscript number (as for normal footnotes). However, VDE (the German association of engineers specialized in electrotechnics, electronics, and information technology) recommends the use of small superscript letters to distinguish the table footnotes from normal footnotes. The table footnote text appears directly under the table without the 4 to 5 cm horizontal line. Create the superscript letters with Format – Character, if there is no function for table footnotes in your word processor.

Tables shall be structured so clearly, that the reader may not slip into another row by accident. To achieve that, you have the following options:

- sufficient line spacing,
- blank lines (after max. five text lines), and
- horizontal lines.

Horizontal lines are by far the safest method to prevent the reader's eyes from losing the current line. Therefore we want to recommend again horizontal and vertical lines to structure tables.

3.3.2 Table numbering and table headings

In the Technical Report tables have a table heading or caption which consists of table number and table title. In the presentation most people only use the table title (i. e. the text part of the table heading). In the **table heading** you may specify, which data is contained in the table, which basic conditions were relevant, which statement or conclusion the table shall proof etc. Moreover, **cited tables must get an indication of source (citation)** here.

The components of a table heading are called as follows in this book:

Table 16 Results of the fuel consumption measurements	table heading
16	table number
Figure 16	table label
Results of the fuel consumption measurements	table title

If the table is continued, the table heading may get a comment, that the table will be continued on the next page. This is useful, because the reader then knows of the continuation right from the beginning of reading the table. The following examples show this and other possible methods of informing the reader about the continuation. In the following examples the complete table is visualised by two short lines. At first an example of a table which expands across one page only.

Table 16 Results of the fuel consumption measurements

Now an example how a table can be labeled which expands across more than one page. On the left there is the table heading on the first page of the table, on the right is the table heading on the follow-up pages:

Table 19 Res... <to be continued>   **Table 19** Res... <continued>

In this example, tables that expand across several pages keep their table number and title on the second page and all subsequent follow-up pages. They differ only in the comment "<to be continued>" and "<continued>" directly after the table heading. If a more explicit labeling is desired, the comment "<to be continued>" can also appear at the end of the table:

Table 19 Res...   **Table 19** Res... <cont.>   **Table 19** Res... <cont.>

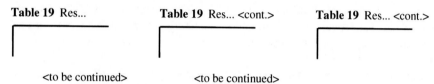

        <to be continued>   <to be continued>

Another way to label the table headings, if the table expands across more than one page, is to apply the numbering method of technical drawings for this purpose. Here is an example using the same table heading as in the last examples.

Table 19 Res... <page 1 of 2>   **Table 19** Res... <page 2 of 2>

The **table numbers** can be consecutively counted through the whole report (example: 1, 2, 3, ... , 67, 68, 69). They can also be a combination of the chapter number and the table number which is consecutively counted within the chapter (example for chapter 3: 3-1, 3-2, 3-3, ... , 3-12, 3-13). Instead of a hyphen you can also use a dot to structure the table number (example: 3.1, 3.2, 3.3, ... , 3.12, 3.13). Books that have many small tables on one page sometimes use a combination of page number and table number within the page (example for page 324: 324.1, 324.2, 324.3).

The consecutive numbering of tables without chapter or page number has the advantage that the total number of tables can be determined quite easily. However, if you want to add or leave out a table late during the writing process, there are predominant disadvantages, because all subsequent table numbers and the cross-references to these table numbers must be changed. If you want to create the cross-references automatically, please refer to sections 3.7.4 and 3.7.5.

It is wrong to write document part number and document part heading into a table header:

       2.2 List of requirements

It is also wrong, to write the document part number and document part heading and then directly under the heading to start with the table. In any case it is better to write a normal document part heading – here **2.2 List of requirements** –, then an **introductory sentence**, and then to start with the table "list of requirements" with the appropriate entries.

If you use Word and you want to **create the list of tables automatically**, you should format all table headings with the same formatting template (style) and use this style only for table headings. Otherwise the automatic creation might be incomplete or might have undesired entries. For details please refer to sections 3.7.4 and 3.7.5.

In general, the table heading shall describe the contents of the table as accurate as possible. Table heading, header, and introductory column shall be understandable without further explanations. The legend and evtl. table footnotes may complete this information.

For the formulation of table headings the same rules are applicable as for the formulation of document part headings, see 2.4.3, **checklist 2-4**. In the text there should be at least one cross-reference to each table which is integrated in the current text chapter. If possible, the table should be arranged near the cross-reference. If there are several cross-references from the text to a table, the table should be arranged near the most important cross-reference. Cross-references from the text to a table can be formulated as follows:

- …, Table xx.   or
- …, see Table xx.   or
- … shows the following table.   or
- … shows Table xx.

Quite frequently the cross-references to tables use the term "Tab. xx" instead of "Table xx".

The table heading shall be positioned near the appertaining table. Therefore the distance above the table heading and below the table should be larger than the distance between the table heading and the table.

If your **table heading** expands **across more than one line**, the layout should correspond to the layout shown in **Figure 3-18 in section 3.4.2** for figure headings that expand across more than one line (the text is left-justified and aligned along a building line).

If you **copy or scan a table from a source of literature** or you integrate a table from the internet as a graphic file into your Technical Report, you should cut off the table number and table title or you should not scan it together with the table. Then you should type your own table number and table title. The table title can either be taken over from the literature source or the internet without change or you formulate your own, new table title. All table numbers and table titles are created in the same way with your word processor. The result is a consistent overall impression.

Tables which you did not develop on your own must get a **note of reference (citation)**. How this note of reference looks like is described in 3.5.4 "Citations in the text" in detail.

3.3.3 The morphological box – a *special* table

The morphological box is – beside its use as a **creativity method** – a central element in **design methodology**. The approach of design methodology is used more and more due to economic reasons. There the morphological box is one work step in the design process. Hence it occurs quite often in Technical Reports.

Characteristic attributes of design methodology are thinking in functions and the sequence of work steps.

Starting from the **list of requirements (requirements specification)**, the main function for the piece of equipment that is to be planned must be defined. It has proven to be practical to set up the list of requirements in the form of a table. Next the **main function** is divided into sub functions. The **sub functions** are always formulated according to the principle **"execution on the object"**. Examples: create force, transform torque, guarantee steerability etc. Then design solutions are developed for each sub function. In the following step the sub functions and the found **solutions for the sub functions** are clearly arranged in a matrix-shape **in the morphological box**, i. e. in a special table. The solutions for the sub functions can be displayed in the table cells only verbally (very often!), only graphically (with principle drawings) or verbally and graphically. From here on two different options how to proceed are possible.

The first evaluation procedure – exactly according to the rules of design methodology in VDI 2222 to 2225 – **concept variants** are defined according to VDI 2222, sheet 1. These concept variants are usually marked in the morphological box with coloured or otherwise distinguishable lines. Please refer to **Figure 3-10** to see such a morphological box with several concept variants. Then these concept variants are evaluated with respect to their technical and economical properties. The results of this **technical and economical evaluation** will be summarized in the **s-diagram**. In the s-diagram you can recognize the best-suited concept variant. Yet, during the technical and economical evaluation the designer has to define, how strong the different technical and economical aspects should be weighted.

In the second evaluation procedure **only one concept variant is selected**. At first a morphological box is arranged without marking concept variants. Then all design solutions of the sub functions are verbally evaluated. This means, that **for all design solutions of the sub functions their advantages and disadvantages are listed in bullet lists**. The design solution that shall be used is announced in the text in a separate line as follows "**selected:** <complete name of the solution of the sub function>". Then follows a short **explanatory statement** of about one to three sentences. The concept variant is now mentally combined and mounted from the best-suited solutions of each sub function. The **marking** of this concept

LIVERPOOL JOHN MOORES UNIVERSITY
LEARNING SERVICES

variant **in the morphological box** follows at the end of the verbal evaluation, **e. g. by shading** the selected solutions of the sub functions **in gray**, see **Figure 3-11**.

If a verbal evaluation of a sub function has only advantages and disadvantages are not specified (or vice versa), then the reader wonders, whether there are no disadvantages at all or formulating the disadvantages has been at first postponed and then forgotten during the writing process of the Technical Report. Therefore it is better – if there are no advantages or disadvantages – to explicitly specify this circumstance during the verbal evaluation of the current sub function. You may express that with the word "none" or with the symbol "–". Now an example for such a verbal evaluation is shown for "Sub function C: lift water".

Solution C2: **Rotary pump**
      Advantages: • smooth
                           • large delivery volume
      Disadvantages: none (or "–")

The morphological box is a relatively complex table, if you compare it with a table of measured values. Therefore, in this book it serves as an example for the design of more complex tables. Naturally speaking, you can also take it as an example in itself, because morphological boxes are relevant for design methodology and occur quite frequently during study courses and in practice.

In the following you will get to know both variants of the morphological box (with several or only one concept variant) in a common example. Both variants of the morphological box refer to a fast turn-off device of a nuclear power plant, which has been developed at University for Applied Sciences Hannover for a design planning task.

In the first morphological box, **Figure 3-10**, several concept variants are marked. In a Technical Report the different concept variants should be marked with colours – if possible –, because the readers can then distinguish the concept variants best. Colored pencil is better suited than other pens, because the colour does not appear on the rear side of the paper, as this is often the case, if you use highlighter or felt pen. Sometimes this even happens with ball pens. Here we used different line types and line thicknesses, because the print is only in black-and-white. In any case you have to undoubtedly assign the different colours or line types and line thicknesses to the different concept variants in a legend below the morphological box.

In the morphological box with **several concept variants** you mark the solution, which is best-suited after the technical and economical evaluation, with a **much thicker line**, if printing and duplication is in black-and-white.

In the morphological box with **only one concept variant** you mark the solutions of the sub functions, which are best-suited after the verbal evaluation, with a **gray shading**, **Figure 3-11**.

Sub functions		Solutions of the sub functions			
		1	2	3	4
A	notice incident	electromagnetic clutch	sensors		
B	disconnect device from normal operation	electromagnetic clutch	hydraulic clutch	pneumatic clutch	
C	lower control rods	electric drive	hydraulic drive	pneumatic drive	self-weight
D	connect brakes	electromagnetic clutch	hydraulic clutch	pneumatic clutch	always connected
E	create braking force	disc break	drum break	induction break	hydraulic damper
F	transmit braking force	cogwheel and cograil	threadrod and nut	friction wheels	
G	control braking operation	centrifugal governor	time controller	distance controller	
H	enable end cushioning	hydraulic oil brakes	hydraulic damper	pneumatic damper	

Legend:
Variant 1 = electric-mechanic solution
Variant 2 = electric-pneumatic solution
Variant 3 = electric solution

Figure 3-10   Morphological box (for a fast turn-off device in a nuclear power plant) with several concept variants

Special Design Rules for the Morphological Box

- The various sub functions in the morphological box usually have a different number of solutions of the sub functions. Since the morphological box is rectangular, there have to be **empty cells** at the right margin. These empty cells **remain white**, they are not shaded in gray or crossed out.
- Entries in the morphological box should **always** be **left-justified** as the entries in the bill of materials and in the list of requirements. If such entries are centred, this results in a too confusing layout.

Sub functions		Solutions of the sub functions			
		1	**2**	**3**	**4**
A	notice incident	electromagnetic clutch	sensors		
B	disconnect device from normal operation	electromagnetic clutch	hydraulic clutch	pneumatic clutch	
C	lower control rods	electric drive	hydraulic drive	pneumatic drive	self-weight
D	connect brakes	electromagnetic clutch	hydraulic clutch	pneumatic clutch	always connected
E	create braking force	disc break	drum break	induction break	hydraulic damper
F	transmit braking force	cogwheel and cograil	threadrod and nut	friction wheels	
G	control braking operation	centrifugal governor	time controller	distance controller	
H	enable end cushioning	hydraulic oil brakes	hydraulic damper	pneumatic damper	

Figure 3-11   Morphological box (for a fast turn-off device in a nuclear power plant) with one concept variant

- If you define several **concept variants** you should give them **meaningful and easy-to-remember names**. Examples:
  - articulated arm, rotatable, and portal roboter or
  - hydraulic, pneumatic, and electric solution.
- In the following parts of the Technical Report these variants are always referred to with their once defined names and even sketches of the concept variants are labelled with these names as well. This is much better than names like Variant 1, Variant 2, Variant 3 etc., because, if you use numbers, you force your readers to often turn back the pages, to see again what were the properties of Variant 3 etc. You should identify your designed solutions of the sub functions horizontally with numbers and the sub functions vertically with capital letters. You should not use both letters I and J, but only I to avoid confusions. That means the sequence is ... G, H, I, K, L ... . This has the following advantages: If you address cell 3.2, it is not clear, whether this is the second cell in the third row or the third cell in the second row. But if you apply the recommendation above, C2 is the second solution of sub function C. This unique name can e. g. be used in the verbal evaluation of the solutions of the sub functions.

- **If a sub function has several subgroups or characteristics**, you can subdivide this sub function in the morphological box. The subgroups are then **"numbered" with small letters**.

  For example, sub function C "store water" (in a water purification plant) shall be subdivided in container number, type, shape, and size. In the morphological box this has to be noted as follows:

Sub functions		Solutions of the sub functions		
		1	2	3
C	store water			
Ca	number	three	five	ten
Cb	type	barrel	canister	bottle
Cc	shape	cylinder with lid	cuboid with handle	cylinder with neck
Cd	size	600 l	100 l	50 l

  The solutions of the sub functions are addressed with a combination of capital letter, small letter, and number, e. g. Cb2 canister.

- If the **solution of a sub function needs to be subdivided into several subgroups or characteristics**, this is not marked with another number or letter, but the term which describes the solution of the sub function expands across several columns and **every subgroup gets its own Arabic number**.

  In the next morphological box two solutions of sub functions are subdivided into subgroups: air cooling and water cooling of a motor. The subgroups each have their own number, so that they can be addressed without doubt with letter and number (B1 ring cooler to B4 flow cooler).

Sub function		Solutions of the sub functions				
		1	2	3	4	5
A	...					
B	cool motor and cylinder	Air cooling		Water cooling		
		ring cooler	tube cooler	forced-circulation cooler	flow cooler	
C	...					

After the design of the morphological box is described, in the next section you will find information regarding evaluation tables.

3.3.4 Hints for evaluation tables

In evaluation tables several variants are evaluated based on different criteria. Examples are tables with criteria for the selection of a location of an industrial company, cost-benefit analyses, or the evaluation of concept variants according to VDI 2222 and VDI 2225 sheet 3. There the concept variants are first evaluated regarding their technical properties, then they are evaluated regarding their economical properties and then two evaluation tables are created.

At first we want to show you the procedure as it is defined in the standards. Then we want to recommend a few deviations from the standard. Please speak with your supervisor or customer in advance, which procedure and table design shall be used.

In VDI 2222 the concept-finding for technical products is standardised, which has the phases planning, concept-finding (list of reqirements), function analysis (specification), concept (concept variants with technical and economical evaluation), draft (assembly drawing), optimisation, refinement (single part drawings), production of a prototype. Beside many other examples the evaluation of a water purification plant is introduced. However, the names are too general. Instead of "plant" you should formulate it more precisely and speak of "water purification plant". The concept variants should get meaningful names, which your readers can keep in mind easily. The header should be emphasized with bold type as here in the book or with gray shading. But now let us look at the individual steps of the evaluation.

Technical evaluation properties of the plant	Points for variants 1 to 4				
	Var. 1	Var. 2	Var. 3	Var. 4	ideal
Blockage risk	2	3	4	3	4
Emission of smell	3	3	3	3	4
Emission of noise	3	3	2	3	4
Required space	1	2	3	2	4
Operational safety	3	3	4	2	4
Sum	12	14	16	13	20
Technical rating x	**0,60**	**0,70**	**0,80**	**0,65**	**1**

Economical evaluation properties of the plant	Points for variants 1 to 4				
	Var. 1	Var. 2	Var. 3	Var. 4	ideal
Excavation	2	3	4	3	4
Concrete work	3	3	3	3	4
Expenses for pipes and fittings	3	3	2	3	4
Assembly costs	1	2	3	2	4
Maintenance costs	3	3	4	2	4
Sum	12	14	16	13	20
Economical rating y	**0,60**	**0,70**	**0,80**	**0,65**	**1**

The strength s of the variants, which results from the x, y-coordinates of the four variants, is now drawn as points s_1, s_2, s_3, and s_4 into the so-called s-diagram. The ideal solution s_i is drawn at the position x=1,0 and y= 1,0. Then a straight diagonal line from the lower left to the upper right corner is drawn that runs across the whole diagram. The best concept variant, here variant 3, is the one which can be found in the far right and far top of the diagram.

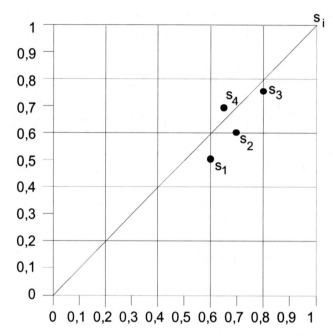

All evaluation tables must give exact information at first sight. If you read the table and the legend it must be clear, which criteria have influenced the evaluation how strong and which variant could gain how many points. Basic rule: The evaluation table **shall not be a mental exercise**. Therefore it must always be explicitly stated which variant has "won": the variant with the most points or – much more seldom – the variant with the least points.

For an appropriate evaluation you often have to **give different evaluation criteria a different level of influence on the final rating** of the concept variants. To express this different level of influence weighting factors have been introduced. You have to multiply the simple point value with the appropriate weighting factor to get the total point value of the variant regarding the current evaluation criterion. Adding the total point values of all criteria leads to the sum total value of the current variant which is listed in the evaluation table in the lowest row.

Figure 3-12 is a concrete example, how an accurate evaluation table looks like. It has been created to evaluate several variants of the chassis of a boat trailer.

Evaluation criterion	Weight	1 axis, 2 wheels		2 axes, 4 wheels		2 axes, 3 wheels, 1 wheel steerable		2 axes, 4 wheels, 1 axis steerable	
	w	SP	TP	SP	TP	SP	TP	SP	TP
Self-weight	10	3	30	2	20	2	20	1	10
Assembly	6	3	18	2	12	1	6	1	6
Price	10	3	30	2	20	1	10	1	10
Ease-of-use	8	3	24	2	16	2	16	2	16
Design	4	3	12	2	8	1	4	1	4
Chassis	8	4	32	3	24	3	24	3	24
Load carrying capacity	8	2	16	4	32	3	24	4	32
Sum total			162		132		104		102

Legend:
w   = Weighting factor   w =   2, very low impact   SP = 0, not suited
SP = Simple points   w =   4, low impact   SP = 1, with deficiencies
TP = w · SP = Total points   w =   6, medium impact   SP = 2, satisfactory
                                w =   8, high impact   SP = 3, good
                                w = 10, very high impact   SP = 4, very good

Figure 3-12   Example of a technical evaluation table for the chassis of a boat trailer

Please look at the legend. At first all abbreviations are explained. Then the meaning of the evaluation factors is defined. So the reader can comprehend, how each "sum total" has been computed. To **avoid logical mistakes**, you should apply the following principles when you give **evaluation points**:

•   The given points are "positive points".
•   High simple point values mean high value and high benefit for the users.
•   Due to the multiplication with the simple point value high weighting factors result in a high level of influence of the current evaluation criterion on the rating result (= the sum total, the added total points) of the concept variant.

Many designers who create evaluation tables have difficulties to give the simple points and sometimes make severe mistakes here! Let us stay with the example chassis of a boat trailer and look at the evaluation criteria self-weight and load carrying capacity: If the load carrying capacity raises, the (positive) point value increases. But if the self-weight raises, then the (positive) point value must decrease! Written like a formula the situation looks as follows:

load carrying capacity ↑   ⇒   benefit for the users ↑   ⇒   simple point value ↑
self-weight ↑   ⇒   benefit for the users ↓   ⇒   simple point value ↓

We can see, that there are obviously two different cases.

<u>1st case: parallel point values</u>   measuring value of criterion ↑ and points ↑
                                                          (here: load carrying capacity)
<u>2nd case: opposite point values</u> measuring value of criterion ↑ and points ↓
                                                          (here: self-weight)

Whether the points must be distributed parallel with the measuring value of the criterion or opposite to it causes many logical mistakes in evaluation tables. It is a problem of language logics which is anyway harder to understand than mathematical logics for many people.

Often several of these evaluation tables follow one another. Then every table must get its own legend, so that every table is readable on its own and unnecessary turning of the pages is avoided.

There are a few keywords signalling "opposite point values". Examples: High effort or expenses is the opposite of high benefit and has to be treated with "opposite point values". Raising *cost* expenditures result in decreasing simple points. A raising *initial training* effort, to learn how to handle a technical product is treated in this way, too. And finally a raising *learning* effort also results in a falling simple point value. When distributing the simple points think of the users or customers and not of the manufacturer or service provider! Then it is easier to decide, whether parallel or opposite point values are required.

3.3.5 Tabular re-arrangement of text

REICHERT has described the method didactic-typographic visualisation (DTV) in several books. In his approach continuous text is broken up, shortened, and visualised. This improves the clearness compared with continuous text. The result are either tabular arrangements of information or text graphics. Here are two examples for tabular arrangements:

Original text:
   The sterilisation temperature for sterilising the tank should be at least 135 °C for 30 minutes. The sterilising temperature at the condensomat should not fall below 125 °C.

By means of DTV improved version:
   Minimum temperatures for sterilising:
   •   at the tank   = 135 °C, 30 min
   •   at the condensomat   = 125 °C

If you draw an outer frame around such a text graphic, there is a smooth transition towards a table. Here another example (some extracts):

Operational disturbances, you may improve or repair on your own

Disturbance	Possible causes	Corrective
Motor does not start	blown fuse	change the fuse
	the thermal overload protector of the motor has tripped	adjust the thermal overload protector
Motor starts slowly	designed for delta connection, but connected in star connection	correct connection
	Voltage or frequency differ much from normal during switch-on	improve power network conditions

After you have got to know several rules and recommendations how to design figures-tables and text-tables, morphological boxes and evaluation tables and after a first short introduction to didactic-typographic visualization, in the next sections you will get hints for creating and designing figures.

3.4 Instructional figures

Information can be displayed in different ways:
- with letters (words, sentences, text-tables),
- with figures (figures-tables and formulas) or
- as graphic display or figure resp. (diagram, illustration, graphic, image, scheme, etc.; a differentiation and definition of these terms will be provided later).

These different ways of presenting information have advantages and disadvantages. The verbal presentation of information is very exact and you can communicate abstract ideas, but it is not very clear. Figures are also very exact, but comprehensive figures-tables are also not very clear, even if you apply an optimal table design. If the figures are visualized in a diagram the clarity is improved a lot.

Many other situations can be explained much simpler and clearer in figures than in verbal descriptions or tables and formulas. **Figures** are an **eye-catcher**, they are noticed first. Information which is graphically displayed can also be read and computed much easier.

Figures create associations in the reader's brain and thus activate the fantasy. Therefore graphics are **motivating and memorable**. Partly they can even substitute detailed text descriptions. Figures **emphasize structures** and visualize the real life. Figures facilitate and intensify understanding processes. Therefore information presented in figures can be **remembered much better** than read information. The efficiency of information transfer becomes better.

However, figures are not always the best way to present information. Abstract information can be better described with text. It is important to find a good mixture and assignment. The author decides, which information shall be displayed as text, which information shall be displayed only in a figure and which information shall be offered in both ways. In this context people use the term text-figure-relationship. This **text-figure-relationship must be well-planned**. In addition, the placement of the figures relative to the text must be planned. By selecting appropriate positions for the figures and by giving cross-references from the text to the figures the author recommends a reading sequence.

The figures can either be integrated in the text or appear in a separate appendix. Both methods of figure placement have advantages and disadvantages. When revising the text the figures can be managed easier, if they are placed in an appendix, since they keep their position in the appendix when text changes are entered. If the figure would be placed in the text, this required additional page layout measures (move paragraphs from "below the figure" to "above the figure" and vice versa, adopting the cross-references from the text to the figure). Placing the **figures in an appendix** has one severe **disadvantage**. Because readers have to turn the pages back and forth **the text-figure-relationship is quite bad and the understandability sinks**.

Therefore we recommend **placing the figures near the appertaining text**. Ideally the figure is on the same page as the explaining text. If front and reverse pages are printed, the figure shall be on the same double-page as the text. If this is not possible due to layout problems, the figure can occur one page later, while a cross-reference from the text on the preceding page to the figure on the next page facilitates that the reader can turn the page at once and evtl. several times.

⌯ *Always give your readers a **timely** cross-reference to figures and tables!*

The easiest method to give such a cross-reference to a figure is to formulate any statement and add the cross-reference at the end with a comma. Example:

> The force flow in the press runs from the plunger via the frame
> to the press table, Figure 12.

If you give such a cross-reference and the figure is not on the same side, the reader will turn the page and expect the figure there. With this very simple hint "..., Figure 12." the text-figure-relationship is assured and the understandability of the Technical Report is improved very much.

⌯ *You should draw or scan your figures as early as possible parallel with writing the text. If you start too late with creating the figures, you might be forced to work very fast (quick&dirty). Every reader can see this in the final product.*

The positive influence of figures in Technical Reports can only unfold well, if certain general rules are kept. These rules aim at good readability.

According to DIN 19045 an illustration shall be simple and well-structured. If possible, the illustration shall show **only one process, item, or thought in one figure** so that the reader can understand it fast and easy.

The width-to height-ratio shall be the same as the width-to height-ratio of **DIN formats** (1 : $\sqrt{2}$ or 1 : 1.41 resp.). Originals of figures, which shall be projected, must be especially well-readable so that the projection on the wall is also well-readable.

To simulate the conditions of the projection look at your original with good light from a large distance – the test distance. According to practical experience a test distance of six times the long edge of the original of the figure is sufficient.

 *For DIN A4 (210 × 297 mm) the **testing distance** for the readability of origi-nals for projection is **at least 30 cm ×6 = 1,80 m.***

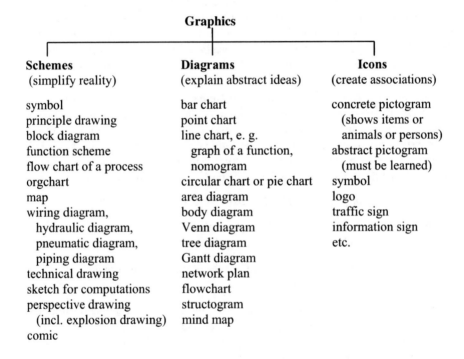

Figure 3-13   Systematic structure of graphic displays according to function and contents

In DIN 19045 part 3 the test distance shall even be eight times the long edge of the original of the figure.

Figures which are not projected, but just printed in a Technical Report should be readable without any problems from the normal reading distance of about 30 to

40 cm. Indices and exponents are often too small, no matter, whether the figure is self-made or copied or scanned.

If a figure, which is to be copied, or a printed self-made figure is checked for readability from the normal reading distance of 30 to 40 cm, suboptimal areas can be identified at once and countermeasures can be applied (larger zoom scale, different font type, larger font size etc.).

If you look at how the message is transported in figures, this leads to the overview in **Figure 3-13**. The further information about figures first deal with **planning** the figures. The basic rules for information-effective design of figures are explained and some rules for figure numbering and figure subheadings are introduced.

Other sections of this chapter deal with the **displayed contents and** the resulting **type of graphic display**. There you can find tips for creating photos, photocopies, digital photos, scans, schemes, diagrams, sketches, perspective drawings, technical drawings, mind maps and pictorial re-arrangements of texts.

3.4.1 Understandable design of instructional figures

Graphics or figures resp. have three possible functions. They shall either **simplify reality** (e. g. principle drawing, map) or **explain abstract ideas** by means of spatial arrangement (e. g. bar chart, pie chart, tree chart) or **create associations** (e. g. logo and pictogram).

In his dissertation thesis BÖHME sorts figures into two large groups: a) motivating and memory-supporting figures and b) informing and problem solving-supporting figures.

If you look at a figure in a possible source of reference, you should decide into which group it belongs and whether it supports your information purpose. If you draw the figures yourself, elaborate the characteristic properties as clearly as possible.

Properties of the motivating and memory-supporting figures
Function: illustrate and help to memorize information, address humor and interest
Impression on the reader: clear, easy-to-remember, inspiring
Properties: easy-to-remember, striking, motivating, humorous, interesting, clever, unusual, provocating, ingenious, with wit and intellect

Properties of the informing and problem solving-supporting figures
Function: explain and complete information in the text, enable understanding of the text.
Impression on the reader: add exactly that information to the text which is shown best in the figure.
Properties: Text contains cross-references to the figures, text guides through the figure, figure has an accurate subheading and evtl. a legend, placement so, that when reading the text you can look at the figure and vice versa.

Figures are often too little exact for these information purposes, but they have instructional advantages due to their similarity with the optically recognized world. In most cases it is especially effective, to present the **overview information in figures** and the **detail information** as text. To prevent misunderstandings and wrong interpretations of your readers, you should keep the following basic rules for figure design, **Checklist 3-5**. If you keep these 13 basic rules, this is already a big step towards "good" figures.

Checklist 3-5   Basic rules for information-effective design of figures

1. **Accentuate important items!**
2. **Delete/leave out unimportant items!** (use max. four to seven graphic elements in one picture, otherwise the picture becomes overloaded.)
3. **Line thickness and font size must be sufficient!** (The figure shall be readable without problems from the normal reading distance of 30 to 40 cm.)
4. The eye follows **dominant lines**. Therefore **relationships of graphic elements** shall be emphasized (lines, arrows, columns, rows, common colour). These relationships should also be specified in detail (What is the character of the relationship? What does it mean?) by means of *labels* in the figure or explanations in the *legend*.
5. **Spatial closeness** of elements *is understood as* **conceptual similarity** (objects near to each other belong together).
6. Elements which are **placed above or below other elements**, are interpreted as **hierarchically superordinated or subordinated**. This emphasizes functional structures.
7. Elements which are placed **beside each other**, are interpreted as **time or logical sequence**.
8. If the elements are arranged in a **circle**, this appears as a cycle, a **frequently repeated sequence**.
9. If one element **surrounds** another element, this is understood in such a way that the **exterior term semantically includes the interior term**.
10. Elements like boxes, bars, lines, columns must be **clearly** marked (either by text labels or by graphical explanations/pictograms).
11. **One type of element** may have **only one function** within one figure or figure series. For example arrows can be used for various types of information: direction of movement, flow of information, cause and effect, note etc. The different meaning of the arrows shall be visible due to a different graphical design of the arrows. So, double arrows are used to express a cause and effect relationship. If in the same figure you also want to draw a double arrow for a moment of torque, this arrow must look much different and the difference must be communicated to the reader.
12. Axes have **large figures** on the (vertical) y-axis **at the top and** on the (horizontal) x-axis **on the right side**.
13. For some diagram types there are **standardized symbols**. Example: DIN 66001 for flow charts defines that a rectangle is used for an operation, a

diamond for a decision and a rounded rectangle for start and end of a procedure. Other standards: DIN 30600, DIN 32520, DIN 66261. Naturally speaking, such standards must be applied.

The application of basic rules 1 and 2 is also called **"didactic reduction"**. **Figure 3-14** shows an example.

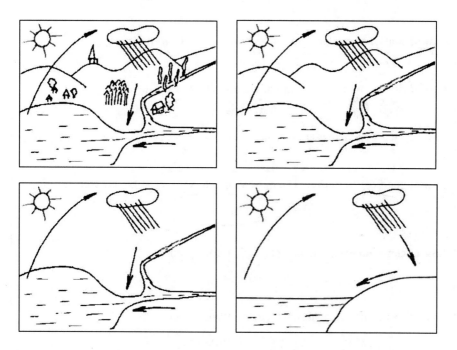

Figure 3-14   Didactic reduction of a graphic showing the water cycle in nature
(source of the first three images: MELEZINEK, Unterrichtstechnologie)

By leaving out and simplifying information the reader can concentrate on the important aspect – the cycle of evaporation, rain and flow back.

The basic rules in **Checklist 3-5** only deal with the global structure of an image. Now we want to discuss detailed mechanisms which are influencing the apperception of figures. Knowing these mechanisms helps to design the figures clearly and without ambiguities even in detail to prevent misunderstandings.

During **optical apperception** the eye scans **individual pixels**. **In the brain** these pixels are connected to shape **lines, points and areas**. These simple geometric objects are then structured and evaluated. The reader recognizes objects or object groups resp. During the interpretation of these simple objects in the brain shape laws (gestalt laws of organization) are applied, **Figure 3-15**.

⊐ *If you use color, you should think about whether your Technical Report shall be published online as well as printed. Colored objects and lines appear much different, when printed in black-and-white, because the printer changes colors to grayscale values.*

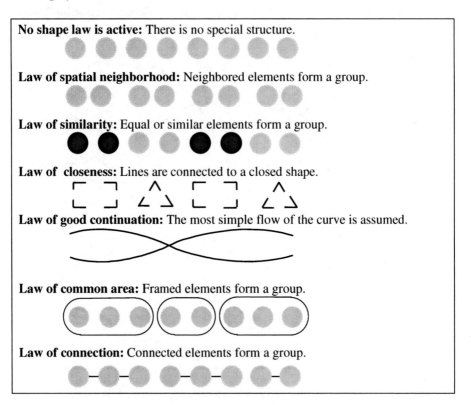

No shape law is active: There is no special structure.

Law of spatial neighborhood: Neighbored elements form a group.

Law of similarity: Equal or similar elements form a group.

Law of closeness: Lines are connected to a closed shape.

Law of good continuation: The most simple flow of the curve is assumed.

Law of common area: Framed elements form a group.

Law of connection: Connected elements form a group.

Figure 3-15   Shape laws apply during the apperception of figures
                    (developed by WERTHEIMER, 1922)

You can influence the order and intensity of the apperception of objects in the figure, if you apply the shape rules and a few other measures. These measures help to accentuate parts of the figure and to influence the reading sequence, **Checklist 3-6**.

Checklist 3-6   Measures to accentuate parts of the figure and to influence the reading
                     sequence

Color: Colors can be memorized much better than section lining, different line thickness, or different line styles. Most favored and striking is the color red. But use color sparing. Otherwise the control effect is lost.

Arrows: There are many graphic design variants available for arrows. Arrows can point out important details. Arrows can also have other functions, see Checklist 3-5, rule 11.

Oversize: Details which are not eye-catching, but important are displayed larger and unimportant details are left out. The scaling factor can be raised up to 1,5 (150 %) without irritating the reader.

Line thickness: Important elements are drawn with 0,75 to 1,50 mm thick lines. Less important details appear with a line thickness of 0,25 to 0,50 mm.

Framing: To accentuate important details they can be framed or marked with a circle.

Detail enlargement: A rectangular or circular frame is repeated in the same figure or (not so good solution) in a second figure (magnifying glass). The connection between the segment in original size and the enlarged segment must be recognizable (connecting lines). The frames must be similar (e. g. rectangle with constant width-to-height-ratio or circle) so that the reader can assign the segment of the figure in original size and the enlarged segment with each other.

Colored background or shading: An important section of the figure gets a colored or gray background. In this case the contrast between object(s) in the figure and background must be sufficient.

Please use only a few of these measures to accentuate parts of the figure and to influence the reading sequence. Otherwise the figure might become confusing. Do not accentuate too many details. Otherwise the influence on the reading sequence might get lost.

If you design your figures or modify figures from literature sources according to the 13 basic rules, keep in mind the shape rules and apply the measures to influence the reading sequence, your readers can recognize and interpret the message of your pictures much easier. Your figures become better understandable. Your message sent out as a figure reaches your readers much better and exactly in the intended way.

3.4.2 Figure numbering and figure subheadings

According to ISO 7144 **figure subheadings** shall be placed **below the figure**. This is correct in the Technical Report, while in a presentation the figure title is placed above the figure. Figure titles are indispensable, but often they are missing!

The components of a figure subheading are called as follows in this book:

Figure 16 Overview of production process variants   Figure subheading
16   Figure number

Figure 16   Figure label
Overview of production process variants   Figure title

The **figure subheading** in the Technical Report (or figure title in the presentation) completes the labels of axes, sectors, bars, etc. and it gives additional information and evtl. **an indication of reference (citation)**. In the figure subheading you should specify what the figure shows, for which conditions it is valid, which statement it shall support etc. The figure subheading should have the following layout:

Figure 16   Overview of production process variants

The figure label is accentuated by bold type printing, the figure title is printed with your standard font.

If a figure spreads across more than one page, you can apply one of the methods to mark the continuation which has been shown for tables in section 3.3.2 analogously for the figure.

The **figure numbers** can be consecutively counted through the whole report (example: 1, 2, 3, ... , 67, 68, 69). They can also be a combination of the chapter number and the figure number which is consecutively counted within the chapter (example for chapter 3: 3-1, 3-2, 3-3, ... , 3-12, 3-13). Instead of a hyphen you can also use a dot to structure the figure number (example: 3.1, 3.2, 3.3, ... , 3.12, 3.13). Books that have many small figures on one page sometimes use a combination of page number and figure number within the page (example for page 324: 324.1, 324.2, 324.3).

The consecutive numbering of figures without chapter or page number has the advantage that the total number of figures can be determined quite easily. However, if you want to add or leave out a figure late during the writing process, there are predominant disadvantages, because all subsequent figure numbers and the cross-references to these figure numbers must be changed. If you want to create the cross-references automatically, please refer to sections 3.7.4 and 3.7.5.

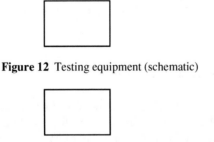

Figure 12 Testing equipment (schematic)

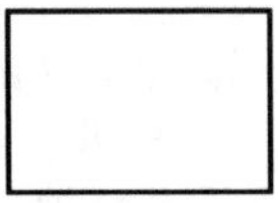

Figure 13 Impeller anemometer for measuring the suction field

Figure 3-16   Bad vertical placement of the figure subheading (especially if there are several figures following one another without intermediate text)

The **figure subheading** must appear **below the appertaining figure**. Besides it should be placed **near the appertaining figure**. Thus, the placement of the figure subheading in **Figure 3-16** is not very clear. The reader asks himself: Where does the figure title of Figure 12 belong (to the upper or lower figure)?

The placement in **Figure 3-17** is much better. There are two (up to three) blank lines between text and figure. The distance between figure and figure subheading is (a half or) one blank line. Below the figure subheading there are again two (up to three) blank lines. Due to the larger vertical distance above the figure and below the figure subheading the reader can easily recognize that figure and figure subheading are a unit.

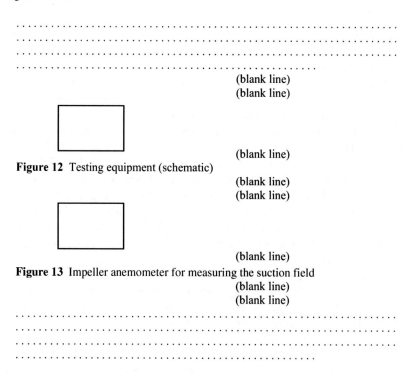

(blank line)
(blank line)

(blank line)

Figure 12 Testing equipment (schematic)

(blank line)
(blank line)

(blank line)
Figure 13 Impeller anemometer for measuring the suction field
(blank line)
(blank line)

Figure 3-17   Good vertical placement of the figure subheading

If a **figure subheading spreads across more than one line**, the second and all subsequent lines start at the **building line** of the figure title, **Figure 3-18**. You can also layout figure subheadings which belong to several small figures like this. The small figures which belong together are placed in a row and marked with small letter and closing bracket.

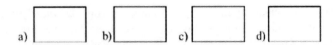

Figure 10 Overview of different methods of welding smoke suction
which is not integrated into the welding process
a) pure hall suction, b) hall suction and hall ventilation,
c) work table suction to the bottom, d) work table suction to the top

Figure 3-18   Layout of figure subheadings that spread across more than one line and
common figure subheading for several small figures

Plan in time, which figures shall be integrated in your Technical Report. It happens that photographs of testing equipment, machines etc. are taken too late or the film is developed too late and due to time pressure the author adds a figure label like Figure 8a between figures 8 and 9 to avoid that all subsequent figure labels and the appertaining manually created cross-references must be updated.

If you want **to create a list of figures automatically**, you should format all figure subheadings with the same (paragraph) style and use this style only for figure subheadings. Otherwise the automatic creation might be incomplete or might have undesired entries. For details please refer to sections 3.7.4 and 3.7.5.

Precise location references are important in figure titles. They help the reader of the Technical Report to imagine the real situation the figure refers to. A figure title with a too imprecise location reference is e. g. "**Figure 6.5** Pressure measuring points P and temperature measuring points T". This figure subheading lacks an addition like "... in the cooling channel" or even better "... in the cooling channel of the die-casting mould".

In this context you should consider that – after a Technical Report has been read once – the readers often only refer to those sections that are relevant for the current problem. Very precise location references in text, figure subheadings and table headings are especially helpful in this case.

 For the formulation of figure subheadings the same rules are applicable as for the formulation of document part headings, see 2.4.3, **checklist 2-4**. In the text there should be at least one cross-reference to each figure which is integrated in the current text chapter. If possible, the figure should be arranged near the cross-reference. If there are several cross-references from the text to a figure, the figure should be arranged near the most important cross-reference. Cross-references to a figure can be formulated as follows:

- ..., Figure xx.   or
- ..., see Figure xx.   or
- ... shows the following figure.   or
- ... shows Figure xx.

Quite frequently the cross-references to figures use the term "Fig. xx" instead of "Figure xx".

If you **copy or scan a figure from a source of literature or the internet**, you should cut off the figure number and figure subheading or you should not scan it together with the figure. Then you should type your own figure number and figure title. The figure title can either be taken over from the literature source or the internet without change or you formulate your own, new figure title. All figure numbers and figure titles are created in the same way with your word processor. The result is a consistent overall impression.

Figures which you did not develop on your own must get a **note of reference**. How this note of reference looks like is described in 3.5.4 "Citations in the text" in detail.

It is quite often the case that you have to **re-number the figures**, because during editing the text of the Technical Report figures are moved to a different section, added or deleted. Then you have to find all figure subheadings in the complete text and all cross-references to figures. This can be executed with the word processor with the function "Find". You only have to enter the search string: "Figure_". The symbol "_" stands for a space character or a tab. Now you can check the figure labels and update them, if necessary.

Even if you plan well in advance, which figures shall be integrated into the Technical Report, it happens quite frequently, that a decision is made rather late to insert a photo of the test equipment, photos of models, specimen or similar figures. The additional work to renumber all the figures can only be avoided by planning early and adding a figure subheading for these photos, which are created later, or you let the computer automatically create figure numbers and cross-references, see sections 3.7.4 and 3.7.5.

There are several alternatives for the **placement of figures: left-justified, centered, or indented by a constant distance**. If the figures are not too large and if this looks balanced, you should select the variant "figure title, left-justified, figures start at the building line at the beginning of the figure title". This has the advantage that the layout is relatively smooth and the figure label is accentuated.

3.4.3 Photo, photocopy, digital photo, scan and image from the internet

Photos on paper and photocopies are treated together, because both have to be **glued** into the Technical Report. Photos are still used today, e. g. for metallographic micrographs in damage analyses, for photos of plants, as press photos and for reproduction purposes in editorial offices of newspapers and journals, book publishers etc. The color intensity and brilliance of paper photos is still higher than of digital photos. However, **digital photos, scanned images and images from the internet** are used much more frequently.

If you want to use **colored images from books and brochures** in your report, scan the image and print the page on a color printer. Later you can reproduce this page of your Technical Report with a color copier. **Colored photos** can be glued into each copy of the Technical Report. The handling of black-and-white graphics with brightness gradients and black-and-white photos is similar.

If you have to create a larger number of copies, you can reproduce the figures with colour and brightness gradients (halftone images) on a black-and-white copier. You should definitely use the photo key. If there are still quality problems even with the photo key, you have to rasterize your figures.

To achieve that you can use **rasterizing film**. The film is laid on top of the original **during the copying** process. The film should only cover the figure, so that the text does not lose its contours and contrast. The rasterizing film disseminates the halftone image to individual pixels. If prepared like this the copy of the page can be used as original for the further reproduction. Rasterizing film for photocopiers is quite expensive, but the copy quality of the black-and-white copies is high. The halftone images can be copied without problems.

Beside rasterizing film for the photocopier you can also get **rasterizing film**, which is used during the **manual enlargement of negatives**. These rasterizing films are exposed together with the negative (a little longer exposure time). The light creates an already rasterized enlargement of the negative on the photo paper. The rasterized pixels in the photo process are finer than the rasterized pixels in the copy process. However, rasterizing film for the photo process is only useful, if you can enlarge negatives manually (i. e. if you have a photo laboratory at your disposal).

If you **scan images or create digital photos**, you should plan early with which **resolution** you want to work. For a simple screen display GIF or JPG files with 72 to 75 dpi are sufficient, while for high-quality journals the resolution must be at least 300 dpi and TIFF files are preferred rather than JPG files.

In the Technical Report 150 dpi resolution is a good compromise in most cases. And the higher the resolution, the larger is the resulting file size.

GIF files can only display a limited number of colors. They are well suited for pictograms and **figures without color and brightness gradients**. Another advantage is that you can combine several single GIF files to an animated GIF file which runs like a small film, e. g. with PaintShop Pro.

JPG files are well-suited for **figures with color and brightness gradients**. If you want to display large images on the screen, the format JPG offers to save the JPG file with the option interlaced. If you look at the image online, the figure is built up different. At first the 1st, 3rd, 5th line etc. of the image are displayed so that you can see a rough outline of the figure quickly. Then the 2nd, 4th, 6th line etc. are displayed so that the image is completely displayed. JPG files do not need much disk space, but it occurs quite frequently that JPG files are blurred, especially screenshots.

Since both formats GIF and JPG have disadvantages, the format PNG is sometimes used. A detailed description of the file format is published on Wikipedia.

PNG is an abbreviation of Portable Network Graphics. PNG has an image compression without losses, it supports different color depths and transparency. The image compression rate of the PNG format is normally higher than the compression rate of GIF files. However, PNG has the following disadvantages: more complex than GIF; no animation; not as high compression rate as JPG, but without losses; no support of the CMYK color model, which is required for four-colour printing, therefore no substitution of TIF files.

No matter which format you use, please make sure that your images are not distorted. **Distortion** happens, if you use a different scaling factor in x and y direction. You can best avoid these problems, if you do not rescale your images with the mouse, but via **numeric entry of a scaling factor**. This factor should result from a multiplication or division by 2 (25 %, 50 %, 200% etc.). Please make sure, that the option proportional scaling or maintain proportions is switched on. (The different word processors and graphics programs call this a bit different, but the sense of the function is always the same.)

When taking photos – no matter whether on classic film or with the digital camera – make sure, that the flashlight does not cause undesired reflections, **Figure 3-19**.

Figure 3-19   Undesired reflection on a photo

🗇 *Try out the required work steps for colored and/or halftone images early enough (test printouts and black-and-white copies of your printouts), so that you can modify the image creation procedures, if the image quality is too low and you still have enough time to keep your deadline.*

In **Checklist 3-7** some rules for creating photos in Technical Reports are introduced.

Checklist 3-7   Rules for the design of impressive photographs

Exposure
- Exposure of slide films oriented towards bright areas ("light").
- Exposure of negative films oriented towards dark areas ("shadow").
- Slide film reacts already to overexposure or underexposure of half an aperture step, negative film starts to react, if the difference is at least one aperture step.
- Therefore: Dias should be slightly underexposed, negatives should be slightly overexposed.
- Digital photos are often far too dark. In interior rooms you should use artificial light.

Picture detail selection
- "Close" to the object or people (leave out distracting information, emphasize details).
- Select unusual perspectives (worm's-eye view, bird's-eye view).
- If the size of the displayed item is not imaginable for all readers without problems, other items should be shown for a size comparison which are well-known to the readers.
  Examples: Ruler, man, hand, finger-tip, banknote, coin etc. If you just specify a measure (e. g. 1 : 25) most readers cannot correctly estimate the dimensions.
- Try out portrait instead of landscape format and vice versa.

Light and shadow
- Light from the side enhances the contrast.
- Reduce backlight by covering the sun.
- Backlight plus automatic exposure without backlight correction results in black objects or people in front of a colored background (only silhouettes).
  That can be a desired result.
- Backlight and a flashlight for the front part of the image result in a harmonic light distribution.
- Emphasis of the front part of the image provides depth (a plastic impression). This is especially nice, if the sides and the top part of the image are emphasized (view through an "archway").
- Accentuations with colors make the image more vivid. Use complementary colors red + green, blue + orange, yellow + violet, but: depending on the target group not too colorful.
- Shaking hands and similar situations should be photographed from the head to the mid of the thigh (above the knee).
- Take enough time (wait for better light conditions, search other standpoints).
- In interior rooms covering light sources with bright cloth or transparent paper and reflecting light with white areas can improve the light distribution.
- The flashlight shall not be directly reflected back from the object or person to the camera.

There are a few rules for modifications of photocopies and scanned or copied figures in your Technical Report. Sometimes these figures are a little overloaded. **Cover not so important details with white rectangles or lines and accentuate important details**, e. g. with arrows or circles. In digital photos and scanned or copied figures you can **weaken not so important parts,** by making them bright and blurry with your graphics program. This is called **"didactic reduction"**, see **Checklist 3-5** and **Figure 3-14**.

To **modify photocopies** you should use a magnifying glass and a drop action pencil with a soft lead, so that you can erase faults easily. Often i-dots, arrowheads and very thin lines (especially section lining) is too small or too bright in the copy. Emphasize them by redrawing! Your modified copy will be the new original and your readers will be grateful, if you invest time and effort here.

If a **copied or scanned image** has reference lines and labels naming important figure elements, there are often **terminology problems**. In the text of your Technical Report you are using one consistent term for an item. Now a comment shall be added about the copied or scanned figure. However, the figure stems from a source which is written on a different linguistic level. Either the source is written too much in general-language or it is too theoretical and uses too many special-language terms which you do not want to use. The situation is similar, if you want to cite from a publication which is written in a foreign language.

Removing an undesired term from the figure is one possible option. However, if the term shall be completely eliminated from the figure, removing the appertaining reference line is not always easy (e. g. when the reference line crosses many closely neighbored lines or cross-section linings). In this case it is better to cover the undesired term with correction fluid or you glue a white rectangle with the desired term into your figure (manually or with a graphics program). It is not so easy-to-read, but acceptable, to explain the undesired term with the desired term in a legend.

📁 *It must be avoided in any case that a figure contains special terms which have not been introduced so far and that there are different names for the same items or procedures in the text and the figure.*

If a figure shall be glued in, there must be enough blank lines in vertical direction so that the figure fits in well. The space left white for the figure can be easily measured with the vertical position information of your word processor (line, column) or with the vertical ruler.

Now you will get some information regarding advantages and disadvantages of figures created with graphics software and CAD programs.

3.4.4 Using graphics software and CAD programs

CAD programs use objects and create vector graphics files. Graphics programs either create pixel graphics files (the image consists of pixels) or vector graphics files (the image is a collection of objects).

If a line drawing is saved as a **pixel graphics file**, it is quite **large**, because for every single pixel the brightness must be saved. However, pixel graphics files are well suited for halftone images (with continuous color and brightness gradients) and for scanned photos. Pixel graphics files are also used in the internet.

Vector graphics files use a different concept. If on a large area a circle is to be displayed, a pixel graphics file would become larger with increasing diameter. The vector graphics file does not change its size very much, because the only data stored in the vector graphics file are centre point coordinates, radius and line thickness, style and color. For a straight line the only data stored are starting and end point coordinates and line thickness, style and color. Therefore these **graphics files** are much **smaller**. Due to the mathematic representation of the figure elements the **objects are scalable**, i. e. the circle diameter or the length of the line can be changed on the screen and printed to paper in different size and at different positions. Therefore vector graphics files are better suited for line drawings than pixel graphics files, **Figure 3-20**.

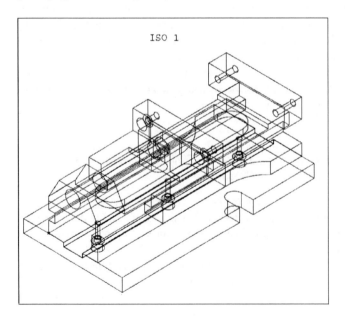

Figure 3-20   Vice designed with a 3D-CAD program (Source: ICEM DDN 5.0 Tutorial)

Using CAD programs saves a lot of time when you modify an existing part, when there are repeat parts and part families. When using CAD programs, the effectivity is much dependent on the task. CAD programs are often too complicated for casual users. They have so many functions, that you have to train and practice using the program. Commercial CAD programs are expensive. In the internet there is a page where they offer several CAD programs free-of-charge: www.freebyte.com.

Programs for creating vector graphics files and CAD programs have partly similar functions. One advantage of many programs for creating vector graphics files and CAD programs is the possibility to create graphics elements on different levels or layers. If you edit such a graphic file you can turn off the levels which are momentarily not needed. Such features are partly also available in pixel graphics programs. No matter which software you use – sometimes it lasts quite long to create PC graphics; but you will be rewarded with good-looking results, **Checklist 3-8**.

Checklist 3-8   Advantages and disadvantages of graphics and CAD programs

Advantages:
- The graphics look **tidy**, there is a **consistent overall impression**.
- By using fill colors and fill patterns area graphics can be created.
- **Changes** can be done **quick and easy**. (You should use a graphics program, which allows drawing on different levels – also called slides or layers.)
- The **graphics can** also be integrated in **word processing or desktop publishing (DTP) programs**. This enhances the Technical Report optically.
- If you use **CAD programs** a perspective **3D display** and the different views can be easily created once the object data has been entered.
- Employers expect that you have learned how to use standard software (graphics as well as CAD programs) during your study course or in an autodidactic approach. Therefore the time you invest in learning to use these programs is not lost.

Disadvantages:
- The **time and effort** you have to invest in **learning complex graphics programs** and for drawing PC graphics is quite **high** (depending on the program). The user interface is sometimes confusing and the time to get first results is so long, that drawing the figures with a simpler program is the better solution for the moment. Pixel graphics programs as well have different concepts for the user interface. If you have to work a lot with pixel graphics (e. g. with images from digital cameras or with GIF and JPG files for the internet/intranet), learn the creation and editing of these files early enough.
- Especially **scanned figures sometimes can have different scaling factors in x and y direction**. And, if you do not mention it, circles are no longer round! It is better to assign the scaling factors numerically.

- If after **scaling** your figure there are **Moiré effects** (checkered or striped pattern), undo the scaling and test scaling factors which are a multiple or a fraction of the original size by the factor 2 (factors 25%, 50%, 200% etc.).
- If after **scanning** a figure with a scanner there are **Moiré effects,** check that with the zoom function of your graphics program. Perhaps it is just a screen display problem. Also, it helps sometimes, to put a thick glass plate onto the scanner and to put the figure to be scanned on top.

An exact estimation how long you need to create digital drawings is normally harder than it used to be when drawing by hand. Therefore we recommend the following:

📑 *Create graphics files as early as possible. Take into consideration that it lasts longer than you estimated, even if it was a generous estimation! In pixel graphics programs using image effects can cause unexpected results. **Test** the functions of your graphics program **on a copy of your data**. Also try out the integration of the graphics into your word processor or DTP program and the printed results as early as possible.*

The next figure is a **vector graphics file created with a graphics program**, which looks attractive due to fill colors and fill patterns, **Figure 3-21**.

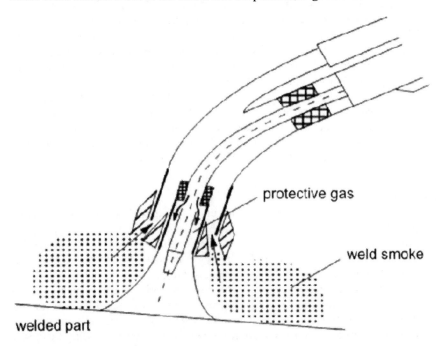

Figure 3-21   Section through a protective gas welding pistol with integrated weld smoke suction (fill patterns and arrows show the direction of flow of protective gas and weld smoke)

⌐ *If the drawings shall look proper even in detail, you should **zoom-in** into tricky areas from time to time and check whether your graphic still looks tidy in the enlargement. Often you cannot see areas which are drawn untidy in 100 % scaling on the screen, but the printer will show everything much more precise.*

Also create **test printouts** of your PC graphic from the word processor program. If possible avoid, that in your word processor or in a graphics program several objects are lying on top of each other in one level, e. g. by using a different graphics program which enables working on different levels (layers). If several objects are lying on top of each other, you have to move the objects lying in the front to be able to select and edit the objects in the back. Then you need to reposition the objects in the front as they were arranged before. Grouped objects sometimes have to be ungrouped and grouped again later etc. All this is time consuming but inevitable without level management.

⌐ *When integrating graphics and scanned images into your text files, you sometimes have to change the **figure size**. You should always enlarge or reduce the size proportionally. In the graphics program the relevant option is called "Maintain proportions" or "Width-to-height-ratio" or "Aspect ratio". In Word you should not "pick" the graphic in one corner and change the size, but select the graphic and use the command Format – Object, tab Size. Otherwise distorted text labels and ovals instead of circles can occur.*

⌐ *The **image resolution** of scanned figures should not be too high. The image resolution for images which shall only be displayed on a screen is 72 or 75 dpi. In the Technical Report an image resolution of 150 dpi is sufficient for proper printouts in most cases. For four-color printing of journals and books it must be min. 300 dpi.*

⌐ ***Image files**, which you use in your Technical Report or in a PowerPoint presentation, should **always be stored on the hard disk drive or a USB stick as separate graphics files**, so that you can enter changes and apply all functions of your graphics program. The image processing functions in Word (toolbars Drawing and Graphics, Microsoft Drawing object, Microsoft Word graphic etc.) and in PowerPoint have less operational scope than the functions of a real graphics program!*

In the next section scheme and diagram are introduced, which occur very often in Technical Reports.

3.4.5 Scheme and diagram (chart)

All images shall be structured as clear and simple as possible. The generally accepted rules of the current field of science and **ISO, EN and DIN standards** must be kept. For several schemes (like flow chart, wiring diagram, hydraulic diagram, pneumatic diagram, piping diagram) appropriate symbols are standardized.

Symbols for flowcharts are standardized in DIN 66001, for technical drawings some relevant standards are DIN ISO 128, DIN ISO 1101, DIN ISO 5456 etc. Other standards defining graphical symbols are: DIN 30600, DIN 32520, DIN 66261.

A program, which is useful for fast and easy creation of diagrams, is Microsoft Visio. With Microsoft Excel you can create diagrams from figures-tables. For simple tasks the operational scope of Word tables and Microsoft Graph diagrams is sufficient. For example, **Figure 3-22** is created with Microsoft Graph.

For diagrams (bar chart, pie chart, curve chart etc.) there is the basic principle that all **axes, bars, sectors etc. must be labeled unambiguously**. Often the understandability of a diagram can be improved by clearly telling in the title above the diagram or in a label next to the diagram which statement the diagram shall support. **Figure 3-22** is an example of how to design a diagram and title or label in a well-coordinated way. The variant with the **diagram title above the diagram** is recommended, if the diagram shall be used for a **presentation slide**. If you write a **text for a brochure,** journal or company report and you use diagrams with titles, but without figure numbers, you should place the **diagram title *below* the diagram without a figure label**. In most cases the title without figure label will be **centered**. However, for Technical Reports it is better to use normal left-justified figure subheadings. Stay consistent, i. e. either all diagrams and figures get figure subheadings (with figure labels) or they get titles without figure labels. The second version is not usual in Technical Reports. Also, the placement and formatting of the figure subheadings or figure titles shall be homogenous.

To make it clear again, the labels of the axes alone are not sufficient to specify the contents of the diagram. The additional diagram title in the figure subheading (in the Technical Report) or above the figure (in the presentation) are indispensable elements of a good diagram.

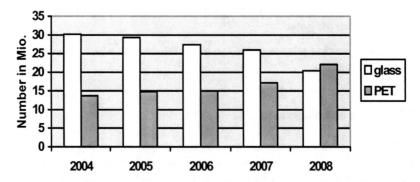

Figure 3-22   Completely labeled diagram with a title specifying the diagram contents
(the diagram is prepared for a presentation)

If a time-related process or development is displayed in a diagram, the horizontal axis is *nearly always* the time axis. **Figure 3-22** is an example for that.

There are additional rules for the display of **curve flows in coordinate systems**. At first, the axes must be precisely labeled with

- **physical value** (as text or formula symbol) and measuring unit,
- **measures** (if it is a quantitative diagram) and
- **arrows** at or beside the axis ends (the arrows point to the top and to the right).

Table 3-1 shows examples for these labeling elements.

Table 3-1   Diagram axes are labeled with the physical value as text or formula symbol and the appertaining measuring unit

Labeling elements	Examples					
Physical value as text	Mass	Time	Length	Force	Voltage	Current
Physical value as formula symbol	m	t	l	F	U	I
Measuring unit	kg	s	m	N	V	A

The axis labeling can appear at the axis ends or at own arrows parallel with the axes, **Figure 3-23**.

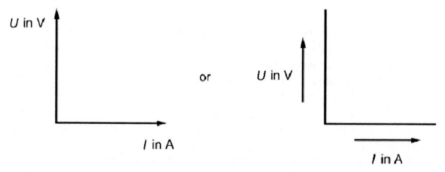

Figure 3-23   Position of physical values in diagrams
at the axis ends or at own arrows parallel with the axes

The physical value can be a formula symbol as in **Figure 3-23** or text as in **Figure 3-24**. A **horizontal text** is much **better readable**, because the graphic must not be turned. If necessary, text of the physical value can be hyphenated.

If you use formula symbols as physical values, the diagram can be used in a foreign language without change. Participants of international conferences can read your diagram much easier.

If you cannot or do not want to avoid a **vertical text** to specify the physical value, DIN 461 recommends **to expand the letters of the vertical text**. However, this might result in a rather long, badly readable physical value text. Therefore we do not recommend expanding the text.

If the physical value is long, it is better to move the whole diagram to the right by a few centimeters and to write a **horizontal text as physical value**, as shown in **Figure 3-24**. In general, horizontal text is **better readable**!

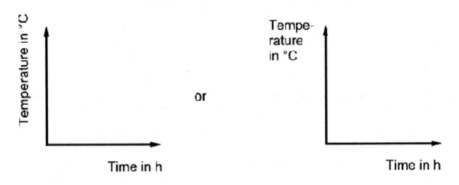

Figure 3-24   Vertical and horizontal physical value texts at the vertical axis
                    (in general horizontal text is better readable!)

If the diagram shall be projected in a presentation or meeting, the horizontal physical value text is the only way of labeling the axes that can be recommended, because your audience cannot turn the figure which is projected onto the wall.

If the **physical value** is given as a **formula symbol** it must be printed in **italic style**, while the **measuring unit is printed in standard style**. There are different variants to specify the measuring unit. In former times they used this method: U [V]. The letter in angular brackets is the measuring unit. However, this method is no longer standardized. Instead, DIN 461 proposes three other options:

- The first variant is to write the physical value at the end of the coordinate axis and the **measuring unit** is written **between the last and the second-last measure** as labeling of the scale, **Figure 3-26**, but this often provides a conflict between font size and available space.
- The second variant is to write physical value and measuring unit divided by a slash, for example U/V.
- The third variant is to **connect the physical value and the measuring unit with the word "in"**, see **Figure 3-25**. This variant is the clearest and is recommended here.

In diagrams with time progression the time must always be the horizontal axis. Proceeding time values begin on the left and go on to the right side.

The **measures at linear scales** (subdivision of coordinate axes) always **include the value zero**. Negative values get a minus sign, e. g. -3, -2, -1, 0, 1, 2, 3 etc. **Logarithmic scales do not have the value zero**.

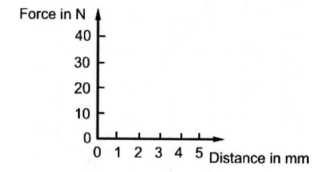

Figure 3-25 Diagram with physical values, measuring units and measures

If a diagram is used to derive measuring values from it, the application of **ruled lines** is useful, **Figure 3-26**. According to DIN 461 the line width of the ruled lines shall have the ratio 1 : 2 : 4. However, this is neither applied in the standard itself nor in widely used textbooks.

In diagrams with ruled lines two factors must be taken into consideration when it comes to defining how narrow the lines shall be: if the lines are too narrow, the reader is confused, but if the lines are too wide, it is very difficult to read off exact measured values. The scaling can be linear on both axes or logarithmic on one axis.

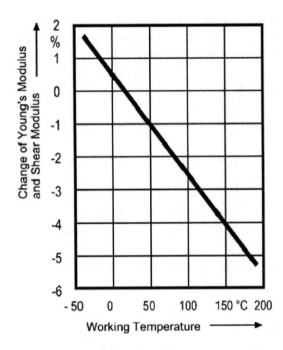

Figure 3-26   Example of a diagram with ruled lines to read off exact values
                   (Source: Table appendix for the textbook ROLOFF/MATEK)

If a diagram is to be drawn with an interrupted scale or interrupted coordinate axis, there are two variants, **Figure 3-27**.

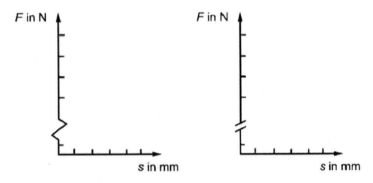

Figure 3-27   Two variants to show that a scale or coordinate axis is interrupted

If **several curves shall be drawn in one diagram**, the curves must have a clearly distinguishable labeling with short, clear terms and labeling letters and figures. The curves themselves must also be clearly distinguishable. To achieve that, you can use different colors, line styles and **measured point symbols**, **Figure 3-28**.

Figure 3-28   Different, clearly distinguishable measured point symbols

The centre point of the measured point symbols is placed in the diagram so that its coordinates equal the x, y coordinates of the measured value. In diagrams with ruled lines the curves and ruled lines can be interrupted before and after the measured point symbols, if this serves the clarity and exactness of reading off the measured values.

If you want to mark foreseeable or admitted **error tolerance zones** in a diagram, there are several variants, **Figure 3-29**.

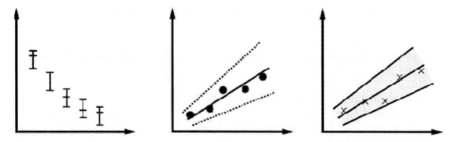

Figure 3-29   Marking of the measured values and the limits of an error tolerance zone within the true value lies (the two left variants give the best contrast)

If you want to get a **marked error tolerance zone** like in the left variant in **Figure 3-29** created by Excel, you can use the following commands:

- enter the measured values into a spreadsheet,
- select Insert – Diagram, select a point diagram type and insert the diagram as a new spreadsheet,
- click on a measured value,
- select Format – Marked data sequence,
- open the tab Error indicator Y,
- click on the preview image Both,
- select further options, e. g. Percentage = 1 % and approve with OK.

Sometimes it is desirable to create a **compensation curve** from your measured values. Excel can do that for you, **Figure 3-30**.

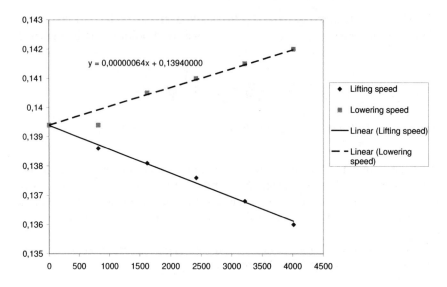

Figure 3-30   Diagram with measured values and compensation curves

The **compensation curve** in **Figure 3-30** was created **with Microsoft Excel** as follows:

- enter the measured values into a spreadsheet,
- select Insert – Diagram,
- on the left side select the type Point diagram, on the right side approve the preset option Points,
- select the area of the spreadsheet, whose values shall be used,
- enter diagram labels,
- select As new spreadsheet and finish with OK.

Now you switch to the new spreadsheet and add the compensation curve:

- select Diagram – Add trend line...,
- on the tab "Type" select the appropriate preview image, here "linear",
- on the tab "Options" select the options *automatically linear* and *Display equation in diagram* and evtl. activate *Intersection point* = (in the example you have to select Intersection point = 0,1394 for the lowering speed), approve with OK.
- Now click on the equation object with the right mouse button and select "Format data labels" from the context menu. Here you should select a larger font size, the figure format "Figure" and 8 decimal places.
- To change the appearance of the trend line, click it with the right mouse button and select "Format trend line" from the context menu. Here you can select, for example, a dashed or a wider trend line.

- Now click on the other objects like labels, data points, diagram area etc. with the right mouse button and select "Format…" from the context menu.

If a diagram shall only show the qualitative and not the quantitative relationship of two physical values, there are no scales on the coordinate axes. However, it is possible to mark important points by labeling their coordinates or physical value (again as text or as formula symbol), Figure 3-31.

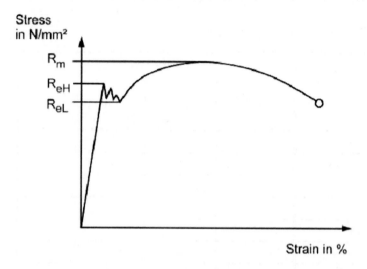

Figure 3-31 Example for a curve diagram that shows only the qualitative relationship of two physical values (stress-strain-diagram of a tensile test)

By selecting a **different scaling density** you can create a **completely different optical impression** using the same data and the same diagram type. In this way you can widely manipulate the optical message of the diagram, **Figure 3-32**. Please make sure, that you do not create an optical impression which represents the original data in a wrong way or which causes misunderstandings.

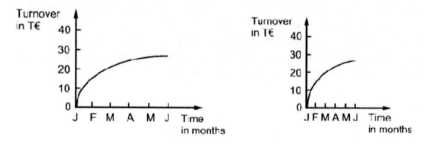

Figure 3-32 Change of the optical impression of a curve by change of the scale density

As already stated – to be sure to avoid misunderstandings the message or conclusion a diagram shall support or prove should be written above or beside the diagram or stated in the figure subheading.

As already stated in 3.3 "Creating *good* tables" – presentation graphics (or diagrams or charts) are very well-suited to display figures in a clear and easy-to-understand manner. The following table by MARKS (slightly modified) shows **which diagram type** can be used **for which purpose, Table 3-2**.

Table 3-2   Diagram types and their fields of application

Information	Diagram type			
development (of time)		(+)		+
distribution (percentage)	+	+	(+)	
comparison	+	+	+	
frequency		+	+	+
functional relationship				+
comparison and development		+		+
comparison and distribution	+	+		
development and distribution	separate charts			

Legend:
  + well-suited
 (+) less well-suited

The standard DIN 32830 "Graphical symbols" specifies **design rules for pictograms, symbols and logos**. It states that these symbols should be built up by basic geometric shapes (squares, circles, rectangles, octagons, etc.). These symbols shall be reproduced large enough, i. e. The width and height should be min. 1/100 of the viewing distance. In DIN 30600 "Symbols, overview" you can find standardized symbols as an inspiration for designing your own symbols.

If the whole Technical Report shall be reproduced by copying, drawing and labeling the figures can be done with drop action pencil.

3.4.6 The sketch as simplified drawing and illustration of computations

When designing technical appliances and machines you also have to deliver a computation of loads. This guarantees that the single parts can bear the applied loads without early failure. In these computations sketches are used to display the geometrical situations, to explain formula symbols and to illustrate results of the computations. If possible, draw all physical values which are computed in your equations. Here we show the following examples: diagram of forces and moments on the arm of a puller, scheme of a gearbox with exact specification of bearings, shafts and gears as well as simplified drawings of a motor, a screw and a cylinder. The sketch as illustration of a computation can also appear as perspective drawing.

Sketches in the Technical Report mostly have **no figure number and no figure subheading**. However, there are rules for sketches, too. They are derived from the basic rule:

⌻ *Sketch for computation of loads = intelligent reduction of technical drawing*

During this intelligent reduction you can leave out details, but never leave out important centre lines. These are absolutely necessary for the quick recognition of part symmetries. Often it is also possible to imitate simplified displays from manufacturer documents.

When you draw machines or plants or parts of them for **handling and manufacturing parts** you should always **draw the handled or manufactured part into your sketch**. For example, it is hard to imagine that in a lifting device for the quantity production of a part exactly this part is missing in the sketch. For example, think of a cask claw. In the sketch for the computation of loads the **cask is drawn in red** and evtl. in a different line style. If the transported part is accentuated like this, the readers can understand the operational processes of the machine, the diagram of forces and moments, the computed loads etc. much better.

Sketches are also used e. g. in **assembly instructions**, to explain the spatial conditions. Here is a text example from an instruction for use of a lawn raking machine by TOPCRAFT, which would be quite hard to understand without sketches (or photos): "Insert bent fastening pipes into case holes. Put anti-vibration rubber disk between case and fastening pipes. Then fix draught relief of cable at fastening pipe. Now the lower push bow is stuck onto the fastening pipe." The reader badly needs the sketch (or photo)!

Sketches for computations of load are always placed near the appertaining computations. They are always **reduced to the essential information**. It is very important to use unambiguous names and labels. If necessary, you should use short explaining texts beside the names and symbols of physical values and points to mark the position in the sketch unambiguously to which the computation refers.

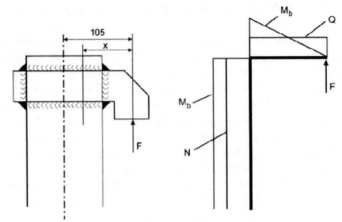

Arm of a puller with diagram of forces and moments

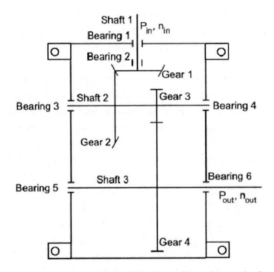

Gearbox scheme with exact specification of bearings, shafts and gears

You may imitate or copy simplified technical drawings, e. g. from **manufacturer documents and standards**, if the clarity is not influenced. Examples:

Electromotor   Hexagon-head bolt   Cylinder

Sketches for computations of load are sometimes **perspective drawings**. Then the rules and tips in section 3.4.7 "Perspective drawing" can be applied accordingly. **If the report shall be duplicated by copying later**, it is also possible, to draw sketches for computations of load **manually with drop-action pencil**. So you can easily correct the sketches.

Eventually figure titles below the sketches are useful to prevent unnecessary searching. For example, if you want to differentiate various profiles:

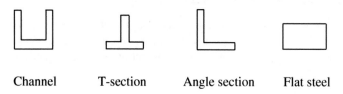

| Channel | T-section | Angle section | Flat steel |

In general, you should apply measures like cross-references, figure subheadings, labels etc. to make sure, that the reader can follow your Technical Report without having to search much and without questions. Imagine your target group as engineers, who have no detailed knowledge of the current project and who shall still understand the report without feedback from you.

3.4.7 Perspective drawing

In technical drawings there are two main kinds of displaying objects. These are the normal three-plane projection (orthogonal projection) and the different kinds of perspective projections. Both kinds of displaying objects are designed in a plane (paper, screen etc.). The difference is, that in the brain the association of the real object is much easier, if the reader looks at a drawing in perspective projection, than it is, if he looks at a drawing in three-plane projection. Thus, the perspective projection is easier understandable.

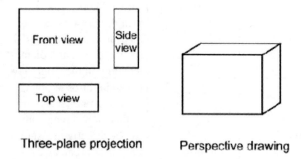

Three-plane projection   Perspective drawing

The following perspective views or perspective projections are distingished in DIN ISO 5456:

- isometric projection
- dimetric projection
- cavalier projection
- cabinet projection
- central projection
  - one-point-method (with one projection center)
  - two-point-method (with two projection centers)
  - three-point-method (with three projection centers)

In mechanical engineering and electrical engineering the central perspective is seldom used. It can be found more frequently in architecture, civil engineering and design. **Table 3-3** shows the perspective projections (all but central projection) in a comparison.

Table 3-3   Comparison of perspective projections for Technical Reports

Projection type	Scale for width, height and depth as well as angle of the edges	Example cube	Fields of application
Isometric projection	W : H : D = 1 : 1 : 1 Edges: -30°, 30° and 90°	90° 30° 30°	Essential information is shown in all three views. Linear dimensions can be directly measured in all three axis directions.
Dimetric projection	W : H : D = 1 : 1 : 0,5 Edges: -7°, 42° and 90°	90° 7° 42°	Essential information is shown in one view. Linear dimensions can be measured parallel with the axes in two directions.
Cavalier projection	W : H : D = 1 : 1 : 1 Edges: 0°, 45° and 90°	90° 45° 0°	Essential information is shown in one view. Linear dimensions can be measured in three axis directions, radii and diameters only in the front view.
Cabinet projection	W : H : D = 1 : 1 : 0,5 Edges: 0°, 45° and 90°	90° 0° 45°	Essential information is shown in one view. Only in this view dimensions can be directly measured. Proportions can be estimated well.

For Technical Reports the **isometric, dimetric and cabinet projection are suited best**. The cavalier projection seems very distorted to the human eye. Therefore it is less suited for Technical Reports. The cabinet projection and the cavalier projection are the easiest to be drawn, since here the projection axes only have the angles 0°, 45° and 90°. These angles can easily be created with a set square (drawing triangle) and they can be drawn easily in freehand technique on normal 5 mm squared paper. You can also use very simple graphics programs and drawing tools to create perspective drawings. Even the drawing tools in your word processor can be used. If you use a simple drawing program, show the grid points and let the geometry snap to the grid points.

In paper shops there are special graph papers and stencils. For isometric projection there is graph paper with millimeter scaling. The lines are printed in the angles -30°, 0°, 30° and 90° (**isometric paper**). Since circles are not projected to all views as circles, there are various **ellipse stencils** and special instruments which facilitate drawing ellipses.

In CAD programs you can create **cutaway drawings,** which – for example – present a look onto ¾ of the case of a drilling machine and into ¼ of the case to look at the gearbox inside. For creating **exploded views** there are specialized (commercial) programs.

Perspective drawings have the following **advantages and disadvantages:**

- clearer, easier-to-understand, better overview,
- saves space compared with three-plane-projection (3 views),
- supports the spatial imagination,
- is more complicated (only cabinet and cavalier projection are simple to create).

3.4.8 Technical drawing and bill of materials (parts list)

The current state of design as well as new developments and changes of parts, plants and processes are documented with texts and technical drawings. Therefore technical drawings have a central function for the communication between technicians. This statement can be expressed in the following more memorable sentence:

🗇 *The technical drawing is the language of technicians*
  and descriptive geometry is its grammar!

Nearly all design and projecting reports have technical drawings, often in a drawing roll or drawing folder. They are an important part of this type of Technical Reports. Therefore we will give you a **list of frequent mistakes in technical drawings**.

Frequent mistakes and questions in technical drawings

Centre lines are often forgotten in technical drawings and in sketches. All parts which are rotationally or axonal symmetric must get centre lines. That is also valid for holes, indexing circles, pitch circles, etc.

If for example holes are regularly arranged at a circular flange, they get a common **indexing circle** and each hole gets a **short centre line**, which cuts the indexing circle in normal direction. The centre lines run perpendicularly to the indexing circle. It is wrong and not according to the standards, to mark each hole with a horizontal and a vertical centre line.

Rotating edges of objects (bearings, bearing caps, etc.) are often forgotten, especially in section and assembly drawings. Therefore, after finishing your drawing, check all rotationally symmetric parts (with the bill of materials), whether all required edges of the objects are drawn.

Joining chamfers are also often forgotten. The remark "all edges which are not especially marked are broken" is not sufficient! Think about how the subassembly or assembly or device can be mounted! Without joining chamfers bearings, shaft seal rings, bearing caps etc. can only be mounted with a larger loose fit and then they usually can no longer fulfill their function. The check **"fictive manual assembly of the parts"** also helps to avoid the additional mistake, that a bearing cannot be mounted, because e. g. a shaft and a gear wheel are manufactured from the solid and the gear wheel is in the way.

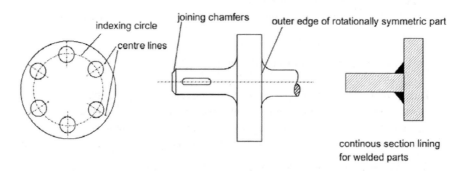

Symbols designating a fit are often filled in wrongly into drawings. The right symbols are: **a small letter and a number for a shaft** and **a capital letter and a number for a drilling**, e. g. ∅ 30 m6 for a shaft and ∅ 30 H7 for a drilling (think of small o fits into large O).

Shafts must be supported in the case (or housing), usually with two bearings per shaft, but the supporting must not be "somehow". The only allowed combinations of bearings are: a) one fixed and one movable bearing, b) floating bearing (two movable bearings with limited possibility to move) and c) angular contact bearing (two angular contact ball bearings or tapered roller bearings in X or O position). However, in design reports of students there are sometimes two sliding bearings or two fixed bearings. That is wrong.

In case of two sliding bearings there is no defined position for gears, pulleys and similar parts. The parts might move and rub at the housing etc.

In case of two fixed bearings the shafts cannot expand when they become warmer. This causes axial loads which are so far unconsidered in the computation of loads for the bearings. Since the load influences the lifetime at least with the exponent 3 (roller bearings: 10/3), such a higher load can shorten the lifetime very much. In extreme cases the shafts bend and the bearings will cant or even jam.

Single part drawings must have **all dimensions** which are required to produce the part. It is quite frequent in **assembly drawings**, that some **assembly dimensions** are forgotten. Here are the most important assembly dimensions: maximum length, width and height (= minimum inner dimensions of the transport container or box), shaft heights, diameter and length of the end parts of the shafts for connecting other parts, index circle or hole distance(s) and hole diameter of flanges to fix the assembly, including flange thickness (because of the required length of the fastening screws, handle lengths, ball handle diameter (where the hand of the user is "connected" to operate the machine or device).

For **assembly drawings of welded parts** with section linings the following rule applies: welded **sub-assemblies are continuously sectioned** (not each plate with a different sectioning), because at the time of assembly the sub-assembly is *one* part.

Frequent mistakes and questions in technical drawings and bills of materials

Now we want to address some problems, which can occur in drawings as well as in bills of materials. According to ISO 7200 in all technical drawings and also in bills of materials the **title block** *must* contain the following information:

- name of the part (or sub-assembly),
- identification number of the drawing, part, sub-assembly, or assembly,
- name(s) of the creator(s) of the drawing,
- date of issue of the drawing and
- sheet number.

The first item is the **part name**. Many engineering students have difficulties here, because the textbooks about technical drawing usually do not give many hints about systematic naming of parts. The task is actually very easy: **Parts are always named "according to their function"**. Examples are axle guide, retaining plate, front-axle pivot pin, stiffening angle, carrying frame etc. In case of standard parts the measures are always part of their names, e. g. Hexagon head bolt M6 x 30. If **any final state of treatment** is mentioned in a drawing or the bill of materials, you always have to use the **participle**: "hardened", "milled" etc.

In the assembly drawing every part gets a reference line and the **position number** from the bill of materials. For the alignment of the position numbers in the drawing and the sequence of parts in the bill of materials there are different approaches. The following scheme has proven to be useful.

Several identical parts get a common position number and are listed in the bill of materials together with their number (part name in singular!). If such a part appears once in several sub-assemblies (e. g. a bearing cap), it is marked at each location with a reference line and the same part number. However, if such a part appears several times in one sub-assembly and mismatches can be excluded, (e. g. four fastening screws of a bearing cap or six columns of a shelves), the part is marked only once within the sub-assembly.

Standard parts appear at the end of the bill of materials in any order. **All other parts** are numbered clock-wise in the assembly drawing. The position numbers within one sub-assembly shall be consecutive and after each sub-assembly two or three numbers remain empty for evtl. later add-ons. The position numbers in the assembly drawing are aligned around the assembly along horizontal or vertical **building lines**.

Regarding the **identification number** of the drawing, part, sub-assembly, or assembly there is the following to say. Classification numbering systems in industry do never consist of one or two digits. Therefore it is wrong, if you create own identification numbers for parts and assemblies consisting of one or two digits. Six digits or more, evtl. with ordering dots, dashes, letters, or space characters are more usual, **Figure 3-33**. The identification number is a mandatory field. It is the only field printed bold in the title block.

The optional field **Responsible dept.** contains the department where the drawing or bill of materials has been created. The optional field **Technical reference** contains the name of a technically "knowing" person who can answer questions. Even if a drawing is created by a company-external consultant, the reference person should be someone who works for the **legal owner** (e. g. the manufacturing company).

During design work in the title block of drawings and bills of materials the mandatory field with the **Created by** must always be filled in. If a group of students has worked together, there are three possible ways to fill in the field. Either only the name of **one group member** is filled in **or an "artificial name"** is constructed from the initial letters of the family names of the group members – evtl. completed by vowels – **or** there is an entry like **"group 3"**. Yet, the last method cannot be recommended, since in case of "group 3" the acting persons are not clearly specified.

Since according to ISO 7200 the field **Approved by** is mandatory, it *must* also be filled in. If there is no real approval process, because students work on a design project during their study course, a thought-out name or the name of a person in the lab who helped much or the name of the supervisor can be filled in.

The next field in the upper row of the title block is empty. According to ISO 7200 it is an **optional field** that can be used for additional classifications, keywords, ordering number of the part for external customers, file name and other organizational information.

The mandatory field **Document type** contains text like: Part list, Assembly drawing, Sub-assembly drawing, Single part drawing, etc. The optional field **Document status** can contain texts like: 1st Draft, 2nd Draft, Translated, Released, Withdrawn. This depends on the release process and regulations from the quality management.

Every drawing and bill of materials must contain the **date of issue** in the title block. In practice it is the date when the drawing or bill of materials or a later revision of it is made available to the public. Therefore fill in the date of issue with day, month and year, **Figure 3-33**. Regarding the **date format**, please use either the format pattern yyyy-mm-dd or yyyy/mm/dd (English and American format). The format dd.mm.yyyy (central European format) can be mismatched with mm.dd.yyyy. Therefore, if you want to use dots to structure the date, it is preferable to write the abbreviated month name in the middle like 12.Apr.2009.

In the field **Rev.** you fill in the revision as capital letter(s) like the columns are named in Excel: A, B, C, … , Y, Z, AA, AB, etc.

Another problem which occurs in technical drawings as well as in bills of materials is the field **Sheet**. To explain the problems occurring here, we should look to the **systematic of technical drawings**. Each technical drawing and each bill of materials (parts list) is uniquely identified by its identification number. According to ISO 7573 the **bill of materials (part list)** and the appertaining **assembly drawing** may be presented as **separate documents**. Then they have different identification numbers and the field **Sheet** is filled in with **1/1 in both documents**. The assembly drawing is free for cross-use and re-use for similar variants. The bill of materials (part list) has a wide form, **Figure 3-34**. The part list and assembly drawing can also be presented as **one document** with a common identification number. In this case there are two options:

a) The part list and assembly drawing are presented on one drawing sheet; the part list has a narrow form and is placed directly above the title block, see **Figure 3-33**. The table header is the last line in the part list, directly above the title block. In the title block the field **Sheet** is filled in with **1/1**.

b) The part list and assembly drawings are presented on several drawing sheets; the part list has a wide form and is sheet 1, the assembly drawing is sheet 2 and evtl. single part drawings follow thereafter with subsequent sheet numbers, see **Figure 3-34**. The table header is the first line in the part list.
The field **Sheet** in the title block has the format <number of current sheet>/<total number of sheets>. For example, if a drawing has five sheets, part list, assembly drawing and three single part drawings, the field sheet would be **1/5** (1st of 5) for the part list, 2/5 (2nd of 5) for the assembly drawing, etc. All five sheets then have the same identification number.

...						
...						
1	10	AB123 001-55	Hexagon head bolt	ISO 4014 – M12 x 80 – 8.8-A2P		
Part ref.	Qty	Part number	Part name	Technical data, designation		Rem

Responsible dept.	Technical reference	Created by	Approved by	<optional field>
CDM-TD	Alan Dempsky	Bart Wayne	David Brown	

<Legal owner>	Document type	Document status
	Sub-assembly drawing	Released
Mieker Getriebebau GmbH	Title, Supplementary title Gearbox housing, complete	217.3567-00
	Rev. A	Date of issue 2009-05-16	Lang. en	Sheet 1/8

Figure 3-33 Bill of materials (part list) with one example entry and title block in the lower right corner of an assembly drawing

Part ref.	Qty	Unit	Reference designation	Part No	Part name	Technical data, designation	Remarks
1	1			26.200-001	Bottom part of housing	EN-GJL-200, EN 1561, R26.200-001	
3	1			26.200-055	Incoming shaft	42CrMo4, EN 10083, Dim Ø55 L=180, ISO 5261	edge-zone hardened
16	1			26.200-167	Oil level gauge	Castrol 27.349.780	
17	1			58.375-081	Pipe, seamless	ISO material, Dim Ø20 1x2,9 L=100	
18	2			99.8765-432	Bracket		
60	2			1234.210	Hexagon head bolt	ISO 4014 – M12 x 80 – 8.8-A2P	
61	24			1234.252	Cylinder head bolt	ISO 4762 – M12 x 40 – 8.8-A2P	black-finished
76	1500	ml		80.12881	Engine oil	SAE 20W HDP 2050	

<Legal owner>	Document type	Document status
	Part list	Released
Mieker Getriebebau GmbH	Title, Supplementary title Gearbox housing, complete	217.3567-20
	Rev. A	Date of issue 2009-05-16	Lang. en	Sheet 1/8

Figure 3-34 Bill of materials (part list) with several example entries presented as a separate sheet in a series with assembly drawing and single part drawings

Frequent mistakes and questions in bills of materials (part lists)

Now some remarks that refer only to bills of materials (part lists). The header of the table is left-justified. The data columns Part ref., Qty and Unit should be right-justified, all other data columns left-justified. The example entries in **Figure 3-34** cover a cast part (bottom part of housing), a milled part (incoming shaft), purchased parts partly with the information where they should be bought (oil level gauge, engine oil) and parts with specification of surface treatment. Please orient yourself along these example entries.

In the column **Remarks** any information can be entered, that might help others to produce and deliver the part or assembly, e. g. a surface treatment like black-finished, nitrided, etc. Such treatments are always specified as participle and de-scribe the end state which is achieved by the treatment. This is also valid for such text descriptions in technical drawings.

In study courses it is usual to prepare the bill of materials (part list) as a sepa-rate document (not integrated in the assembly part drawing). The bill of materials is integrated into the Technical Report *and* added to transparent originals or plots of the drawings in a drawing roll, drawing folder, or drawing box.

3.4.9 Mind map

Mind Maps are used for **structuring topics, problems, plans, discussions** etc. For example, you can draw a mind map during a presentation or lecture instead of writing the usual continuous text script. Mind maps also support **brainstorming processes**. In a mind map all aspects of a topic that must be considered become visible as main branches, which are branched to smaller twigs and in the end have detailed topics, questions, or aspects as leaves.

On the market there are many **computer programs** to create mind maps very fast and easy. For the following example, **Figure 3-35**, the quite price worthy pro-gram **Creative MindMap** by Data Becker was used. In the mind map branches and twigs can be created quickly and a branch can be moved together with its twigs to a different location of the topic tree etc.

An interesting feature is the export of the tree structure to a txt file, which can be imported into Word and prepared for PowerPoint. Every aspect which is going to be a slide is a simple paragraph, the sub aspects are formatted in Word with the automatic bullet list feature and ready you are. Now you can transfer the text doc-ument to PowerPoint with File – Send – Microsoft PowerPoint. You get one slide for each aspect or topic in the txt file, the sub aspects are already formatted as list items in PowerPoint. Now you can visualize your slides with graphic objects, structure them by use of color, add images etc., see 3.8 and 5.4.1.

More expensive and more professional software like **MindManager** have more comprehensive clipart libraries and more features.

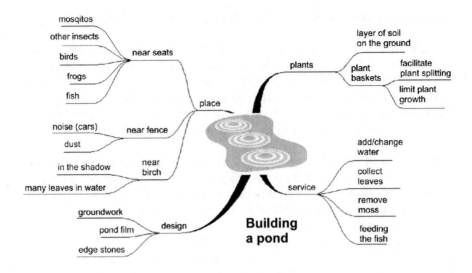

Figure 3-35   Example of a mind map for planning the building of a pond in a garden

3.4.10 Pictorial re-arrangement of text

REICHERT has described the method didactic-typographic visualization (DTV) in several books. In his approach continuous text is broken up, shortened and visualized. This improves the clearness compared with continuous text. The result are either tabular arrangements of information or text graphics.

A text graphic consists of text, which is layouted with typographic measures like indentations, bullet lists, bold print etc. and graphical elements. These graphical elements are, for example lines, rectangles or circles, which are arranged "above" the text. The result is a smooth transition towards a diagram or chart. It is sufficient to use only a very limited number of these graphical elements to create a text graphic from a small amount of text:

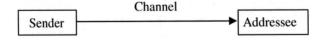

In this book this method has been applied quite often, e. g. in subchapter 2.1, Checklist 2-1 (General overview of all required work steps) or in section 3.1.1, Figure 3-2 (placement of information on cover sheet and title leaf) or section 3.3.3 (morphological box with several concept variants) or in subchapter 3.4, Figure 3-13 (Systematic structure of graphic displays according to function and contents) as well as Figure 3-15 (shape rules) or section 3.4.4 (citation).

To arrange lines, rectangles, circles and other graphical objects "above" the text in Microsoft Word, you have to open the menu Special – Options, tab General and switch off the option "Automatic creation of a new drawing area when inserting AutoForms". Here are some examples as an inspiration for your own text graphics:

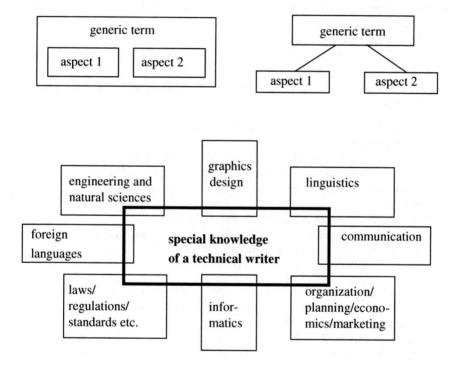

The graphical objects can be roughly positioned with the arrow keys or the mouse, the fine positioning is possible with Ctrl and the arrow keys. The text is roughly positioned with tabs, the fine positioning is possible with space characters.

3.5 Literature citations

When writing Technical Reports there are various versions of literature citations and lists of references. Therefore it will be described in detail how to cite literature and to write a list of references. But here are at first a few introductory remarks.

3.5.1 Introductory remarks on literature citations

These introductory remarks shall begin with a definition of the term "literature citation". A citation is using a statement written by another author either literally or in one's own words and precisely documenting the literature source from where the statement is cited. The short note in the text, where the literature comes from is called citation. In the back in the list of references the cited publications are listed with their bibliographical data.

In scientific works there are nearly always literature citations. Also in Technical Reports, which are partly not so strictly following the rules, there are often citations of statements written by other persons. **Literature citations** have the following **tasks**:

- they help to describe the current state-of-the-art,
- they support the author's opinion,
- they emphasize the scientific character of the Technical Report,
- they exculpate the author from the responsibility for the contents of the citation (but not from the responsibility for the selection of the citation),
- they point out the author's accuracy and intellectual honesty,
- they underline the author's authority and credibility and
- they permit that the readers can check the facts (statements, values in computations etc.) and by studying the literature from several authors to get a higher level of expertise in the current field of science.

Presenting the bibliographical data of cited literature helps the readers to borrow the literature from a library or buy it from a book shop. Therefore there must be sufficient information for each publication.

If every author would use his or her own **systematic** to present the citations and the required bibliographical data, there would be a chaos. Therefore there are two standards, that define how literature citations must look like: **ISO 690** "Documentation – Bibliographic references – Content, form and structure" and **ISO 690-2** "Information and documentation – Bibliographic references – Part 2: Electronic documents or parts thereof".

3.5.2 Reasons for literature citations

Correct citations are a proof that the author knows the rules of scientific working. The selection and quality of the literature citations document how intensively the author has read about the "state-of-the-art" and the established theories. Thus, literature citations do not express, that the author has not had own ideas. On the contrary, it is absolutely necessary to make correct literature citations to facilitate that interested readers can get access to a field which might be new to them.

Correct literature citations emphasize the **honesty** of the author. it is part of a positive human and scientifically correct behaviour, that literature citations are marked. If you copy texts or figures from other authors into your Technical Report without naming the literature sources you pretend that the work results of other authors are your own. That offends national and international copyright laws and good manners within the scientific community.

Someone who **knows** the field well, in which you have worked, will quickly recognize the spots where you have used thoughts of other authors, but without correct literature citation. No one wants to come into such a situation. Therefore the next section is a short description, what information is contained in ISO 690 and ISO 690-2. Then there are sections describing how literature citations must be made in the text and in which form the bibliographical data must be listed in the references. A further section deals with literature written in foreign languages. At last there is a section about copyright law: What is valid in scientific working, in universities, and in teaching situations? What is valid for companies and private persons? What can be the results of missing or incorrect literature citations?

3.5.3 Bibliographical data according to ISO 690 and ISO 690-2

ISO 690 and ISO 690-2 contain rules for collecting lists of references. They describe which bibliographical data must be listed for the cited publications in which order. The layout of the list of references is in block format – a very compact form, which is used in many books and journals. In the standard there is no information how the list of references is presented in the classic three-column form which is usual in Technical Reports.

The term **"publication"** in a narrow sense contains textbooks, contributions to host documents, articles in journals, company literature, brochures, catalogues, etc. These cases are covered in ISO 690. In a wider sense the term "publication" also contains the following media: record, radio transmission, video or TV film, computer program, documents which are saved on CD-ROM or available via internet or intranet. These cases are covered in ISO 690-2. A last case not covered in the ISO standards is a personal message. In this book we want to call all these publication types "literature" and "source".

The next two sections show how the citation in the text has to be made and how the list of references must look like.

3.5.4 Citations in the text

How a citation is marked in the text, depends on what is cited and whether the citation is literal or analogous. We have to distinguish:

- literal citation of text
- analogous citation of text
- exact copy of a figure or a table (scanned, photocopied, redrawn)
- analogous citation of a figure or a table (adopted to the own information target by means of modifications)

Every publication from which information is cited must get a **citation in the front in the continuous text** and an entry with all **bibliographical data in the back in the list of references**. It is not allowed to copy five articles from journals, put them into a plastic pocket, and label the pocket with "Brochure 5", if the five articles thematically belong closely together, but have been published in different journals and written by different authors. It is also not allowed to type a citation directly behind a document part heading and to hope, that the readers will understand that!

Basically, all literature should be **at first listed in the back in the list of references**. in the **next** step the cited information **in the front** should be typed **into the text**.

For large projects with intensive literature work the following work technique has proven to be practical: For all cited publications you should not only note the bibliographical data, but also unique identifiers (ISBN for books, ISSN for journals, SICI for articles in periodicals, DOI for electronic publications, LCCN for publications registered by the American Library of Congress, ordering number of the publisher etc.), the library where you found the publication, location within the library, and signature.

Now you can give back the literature and evtl. borrow it again later, if you need to improve your text or forgot something in the list of references. So you do not need so much space for the books, and you do not need to extend the loan period after four weeks. If you borrow the books again, you have all required data available without having to search the books again. That saves much work and evtl. reminder charges.

Every literature citation consists of the cited information and the citation in the front and the bibliographical data in the back in the list of references.

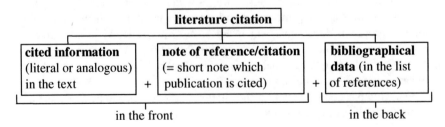

When you cite text there are some rules, which are listed at first and then explained with examples. We introduce variants. Please select one variant and apply it consistently throughout your Technical Report!

Rules for citing text

1. The short note near the cited information in the text shall enable the readers to identify the corresponding entry in the list of references or in a footnote or endnote without doubt. Possible variants are listed below, examples are marked in bold and italic print:

Basic referencing schemes:
- number or symbol written in superscript, referring   *cited info[18]*
  to a footnote or endnote (**= note referencing**)
- own literature number in rounded or square   *[18] or (18) or*
  brackets (**= parenthetical referencing**)   *[Miller, 1998]*

Short citation:
- only the own literature number   *[18]*

First part of long citation (author part):
- author name   *SALINGER*
- or author name and first name(s)   *WINTER, John*
- abbreviated first name(s)   *HENSON, D.*
  STIEG, MF.
  HALDANE, JBS.
- two or three authors   *MILLER, P. and*
  JOHNSON, D.
- four and more authors   *MILLER, P. et al.*
- not mentioned in the standard:
  several authors with identical first names   *KLARE, Donovan I*
  can be distinguished with Roman numbers   *KLARE, Donovan II*
- If a publication has no named author,   *DIERCKE atlas of*
  a short name of the publishing institution   *the world*
  with the location of the head office in brackets   *BYU (Provo)*
  or one or more words from the title
  are used as author name
- The letters N. N. are not used according to ISO 690. *Publisher unknown*
  If not only the author, but also the publisher is
  unknown, a phrase can indicate that.
- optional: specification of the author's function   *MILLER, Peter (ed.)*
  WARNCKE, Tilo
  (interviewed)

- optional: author name(s) in CAPITAL LETTERS
  or SMALL CAPS is more distinctive
  compared with the continuous text

Second part of long citation (date part):
- year of publication   *2003*
- several cited publications of one author
  in the same year are distinguished with small letters *2003a, 2003b, 2003c*
- estimated year of publication   *ca. 1920*
                                                                         ca. 1920 (copyright)
                                                                         ca 1920 (printing)

Third part of long citation (location part):
- optional: page number(s) of the source   *pp. 27-30*
- optional: specification of a document part   *subchapter 9.3*

The citation is enclosed in round brackets according to ISO 690. In Technical Reports it is often enclosed in square brackets to distinguish literature from formula numbers. Square brackets are recommended by several institutions relevant in natural sciences and engineering like ACS, AIP, AMS, CSE, ASME, and IEEE. Here are some examples of complete citations in the parenthetical referencing scheme:

[18]
[KUHN, 1986a] etc.
[18, p. 50]
[HAMSING, 1993, p. 27-30]
[HAMSING, 1993, section 2.7.3]

In the examples above two methods of citing literature have become clear:

a) only specification of the literature source (in the note referencing scheme and also in the parenthetical referencing scheme, when you specify only the literature number)
b) specification of the literature source with the year of publishing and with or without the exact location in the source (page numbers/document part)

Method a) is often used in Technical Reports and according to ISO 690 it is correct. Method b) is nearly always obligatory in the humanities. Method b) is called author-date system or Harvard referencing style. If there are exact page numbers or document part numbers, *all* parties concerned, i. e. the author himself, his supervisor, and the other readers can find the cited information faster.

There are various ways to sort the literature sources in the list of references:

- **Numerically by their order of appearance in the text**
  The first cited source gets the number [1], the second source the number [2] etc. Then the list of references is not sorted alphabetically. This way of numbering is suited, if in the document there are only a few sources cited, e. g. in short Technical Reports or in journal articles. It is e. g. recommended by IEEE, ASME, and ICMJE.

- **Numerically by the alphabetical order of author names**
  In larger works the list of references in the back is in most cases sorted by the alphabetical order of author names. The literature sources do not appear in the text in the numerical order [1], [2], [3] etc., but e. g. in the sequence [34], [19], [83] etc.
- **Alphanumerically by the alphabetical order of author names, literature identifier contains first letter(s) of author name and number**
  This method is not listed in ISO 690, but the ordering in the form [M1], [M2], [M3] etc. or [HAM93] saves time and effort in larger Technical Reports and it is recommended by AMS. If a source of literature is added or left out late in the writing process, only those literature numbers with the same letter(s) must be changed, not all subsequent literature numbers. In the end the alphanumerical identification *can be* replaced by a numerical identification.
- **Please contact your supervisor or customer and arrange which method is to be used!**

2. In large alphabetically sorted lists of references there is sometimes the question, **which is the alphabetically correct order of the author names**. The following rules apply:

   - Several authors with the same surname are ordered by their first names.
   - Several publications of the same author are sorted by the year of publication (2008, 2006, 2005). If the citations do not only contain the literature numbers, but author and year or author, year, pages, identical years get small letters to distinguish the publications (2006a, 2006b, 2006c, ...).
   - Several publications of the same author with several co-authors are sorted by co-author names.
   - If the co-author names are identical, but their first names differ, co-authors are sorted by their first names.

   These sorting rules are applied accordingly, if there are even more identical sorting attributes, until a useful sorting attribute is found. Not mentioned in ISO 690: If there is no useful sorting attribute, use Roman numbers as a distinction.

   - Author names with "Mac..." are treated as author names with "Mc...".

3. Citations from a primary literature source (citation from original work) must be specified as listed in 1., e. g. in the systematic "author, year, pages". If a literature source is hard to get or not available at all, but cited from another author, whose work you have, you can make a **secondary citation**. The citation must look like this:

   [author (*not available*), year, pages cited by: author (*available*), year, pages]

   Both publications are listed in the list of references.
   Naturally speaking, if the secondary literature source only cites author and

year of the primary literature source, you can only copy these two pieces of information.

> In the front in the text there is e. g. the following citation: [KLARE, 1963, 1974/75; TEIGELER, 1968 cited by BALLSTAEDT et al., 1981, S. 212]
> In the back in the list of references you have to list the three primary sources by KLARE and TEIGELER and the secondary source by BALLSTAEDT.

4. After **document part headings** there are never citations.

5. If a single **literature citation is long**, that means it has several paragraphs and – beside the quotation marks – cannot be distinguished from the normal running text, the citation must be **repeated at the end of each paragraph of the cited text**. If the citation stood only once at the end of the last paragraph of the cited text, this would be an uncertain situation. The wrong conclusion would be possible, that the citation only refers to the last paragraph. If you mark the literature citation by italic printing or an indentation, you can write the citation only once at the end of the cited text.

6. Each **literally citied fact** is marked with **quotation marks**.

> "The ongoing technical development is one reason for the rising prosperity in the industrialized countries." [HERING, 1993, p. 1]

7. If you want to cite **only a part of a sentence** or you do **not cite a few words in the middle of a sentence**, this must be marked with an **ellipsis**.

> "Therefore written queries ... have been sent to all professional associations of technical writers in the world and to INTECOM."
> [HERING, 1993, p. 338]

According to DIN 5008 the ellipsis consists of three dots without parentheses. According to Wikipedia the three dots (ellipsis) must be enclosed in angular brackets [...] in citations, if the author left out one or more words.

8. If in the source text there is a **typographic accentuation**, you have to **copy** that in a literal citation **exactly**. This holds true for bold print, italic print, use of capital letters, indentations etc. In the example the title of an article is literally cited and the italic print is used exactly as in the original article. If the italic print would have been omitted, the result were a distorting mistake.

> DILLINGHAM writes in his article "*Technical* Writing vs. Technical *Writing*", that the knowledge of the technology is necessary for technical writers, so that they understand the products, they shall describe. This understanding of the technical details is the prerequisite of successful user information. [DILLINGHAM, 1981]

9. If you do not want to copy the typography of the source or give **comments regarding the cited text**, you have to inform your readers, that you have chan-

ged the typography or presented your personal opinion. The easiest way to do so, is to use the text "note from the author:" and enclose the note in **angle brackets <...> or angular brackets [...]**.

... <note from the author: typographic accentuation in the source text was not copied>.

10. If in the **original text** there are **quotation marks**, this must appear as **half quotation marks** ‚...' in the literally cited text (in Word: Shift + ' [above #]).

"In this context it is especially remarkable, that he <note from the author: Leonardo da Vinci is meant> has introduced a completely new kind of visual display already around 1500 to the technical documentation. This is the 'explosion drawing', ..., which can be found e. g. in nearly all current vehicle spare parts catalogues."
[HERING, 1993, p. 18].

11. **Information that is cited analogously is not marked with quotation marks**. It is only marked with the citation at the end of the cited text:

WIERIGER assumes, that the gas drive is more ecological [12].

12. The **citation** can appear at **different positions within the sentence**. Preferably it should be integrated so that it disturbs the reading flow as little as possible. However, if misunderstandings are possible, where the cited information comes from, the citations must be placed so that the relations become clear.

The physical basics have already been examined by SIMON [17]. He found out, ...

This conclusion could be affirmed by physical [12, 17] and chemical [9, 22] experiments.

Similar research [2, 7-10, 15] shows, ...

Taking into consideration the findings of SMITH [16] and RIEMERS [9] ... .

13. The marking of citations with **angular brackets** "[   ]" is often used in textbooks and similar publications and we recommend it here. If you use round brackets "(   )", like proposed by ISO 690 and often used in the Anglo-American language area, the short citations (only using the literature numbers) could be mixed up with equation numbers.

14. If **short citations** are applied (only using the literature numbers) like "[23]", this citation **should not stand alone on a new line**. Try to shorten the text before the citation so that the citation jumps up one line or add a few words before the citation.

15. **Data** and information, you use in **computations of load**, **error calculations** etc., must be cited. Examples: physical constants, material constants, computation procedures by manufacturers, standardized measuring procedures.

Rules for citing tables and figures

1. If you copy a figure or a table from a source **without changing the contents**, there are three ways of presenting the citation. As seen for citing text, you can type the normal citation. That is, you specify the number of the literature source in the list of references or the author's name, year of publication and evtl. page numbers or section. Again, we recommend square brackets. In any case, the citation appears behind the figure subheading or table heading. Quite frequently the note "(Source: author, year, pages)" with rounded brackets is used. If you have enough space, it looks more balanced to write this note left-justified with the rest of the figure subheading or table heading on a separate line, see the example "Figure 23" below.

   In case of figures from brochures or corporate publications or company homepages you can also use the note "Corporate photography:" plus company name. Examples:

   Figure 13   Section through an Otto motor [15, p. 50]
   Figure 15   Different combustion chamber geometries [18]
   Figure 17   Fuel consumption b_e with different combustion chamber geometry and equal cylinder capacity [MILLER, 1990, S. 35]
   Figure 19   Octane rating depending on Benzol concentration [MEHRING, 1992a]
   Figure 23   Percentage of lightweight construction materials in self-weight of formula 1 cars from 1950 to 1995 (Source: LEHMAN, 1995, p. 81)
   Figure 28   Different designs of needle bearings (Corporate photography: FAG)

2. If you copy a figure or a table from a source with changing the contents, you should use the note "acc. to <author>" or "according to <author>" to express that it is an analogous citation. Your changes could be the following:

   • leaving out/changing labels, so that they fit to the terminology of the text,
   • leaving out details from the figure, and
   • leaving out not so important rows or columns from a table.

   Figure 13 Viscosity of different engine oils depending on the temperature according to PFITZNER [12]

   Alternatively, you can give your readers a note in the figure subheading or table heading like "…, simplified compared with the source" to make clear, that it is not an exact copy, but that you have changed, left out, improved, or further developed something.

3. If you use an **image from the internet**, you have to cite it correctly. We recommend, that in the last line of the figure subtitle you cite the internet address of the image (a) or of the HTML page on which the image is used (b). The examples show a result of a google image search. The internet address has to be entered to the list of references as well. There it has to appear together with an author and the date when the citation was recorded/when the figure was seen. **If the internet address is too long**, you may shorten it so that the reader comes near the relevant topic on the cited homepage. Then please specify, where the reader has to click to reach the cited page. Or you write a shortened address or just the start address of the homepage in the figure title, e. g. http://www.physics.sjsu.edu/becker/physics51/, and the complete address of the figure in the list of references, e. g. http://www.physics.sjsu.edu/becker/physics51/images/25_02ParallelPlateCap.JPG.
    Also, if a **HTML page uses frames** (you can recognize that when the internet address stays the same after you clicked a link, but the contents of the page has changed), please specify, where the reader has to click, to reach the cited page. Examples for literature sources from the internet:

> **Figure 25** Working principle of a parallel plate capacitor
> (Source: www.vtf.de/p90_1_3.gif)   (variant a)
> **Figure 25** Working principle of a parallel plate capacitor
> (Source: www.vtf.de/p90_1.shtml)   (variant b)

☞   *The same rules for specifying the internet address of the files you found apply for other file types like PDF, Word, or Excel files, MP3 files etc.*

☞   *Please consider that the presented contents in the internet can change very quickly. If you found a valuable source, you may wish to copy the texts and figures, which you might use later, to your hard disk drive and note in a file, where and when you found the information (i. e. URL/URI and date, when you saw the information).*

3.5.5 The list of references – contents and layout

In the list of references the bibliographical data of the cited publications is collected. The list of references is always located directly after the last text chapter, normally after „Summary and conclusions". The numbering of the chapter "References" is according to ISO 2145, see also 2.4.2, 2.4.4, and 3.1.2. In articles there are sometimes just intermediate headings without document part numbers. These headings are printed in bold type and usually with a slightly larger font size. The heading for the list of references is also "Works cited" or "Literature" or "Bibliography".

It might be useful, especially for larger Technical Reports to put up a **structured list of references**. In the structured list of references different types of literature are separated by intermediate headings, see chapter 7 in this book. The easiest way to find the structure is to use a heading for

- standards and regulations (ISO, EN, DIN, VDI, VDE etc),
- links to the internet,
- books and articles.

The intermediate headings may start with the term "Used ...". You may also sort the list of references by topic. Then there could occur titles like "Literature on building construction", "Literature on underground construction" etc.

At the end of a structured list of references there is sometimes a section "Further links to the internet", "Further literature", "Further reading" or "Bibliography". Here you can list sources of literature, which have not been cited, but which are important for the treated topic: standard text books, literature on special topics, homepages of companies etc. The section with the bibliography can get an own document part number or – better – just a heading which is emphasized by bold printing and a slightly larger font size.

In very large Technical Reports you can also find the variant of a space-saving **capitular list of references**. The capitular list of references is located at the end of each chapter. The corresponding subchapter headings might be "1.7 References in chapter 1", "2.9 References in chapter 2" etc. The layout of capitular lists of references is similar to the layout of the list of references in journals. An example can be found below under "space-saving form". In each chapter the numbering of the sources of literature starts again with "[1]".

Before you enter your first source of literature to your list of references, you have to define, which layout you want to use. There are three options: the first one is the classic **three-column form**, the second is the two-column **space-saving form**, which is also used in journals or third the **block format**. In the block format according to DIN 1505 the long citation, which occurs in the text, is repeated and printed in bold. Then the bibliographical data is listed. That has the advantage that it is nearly excluded, that the readers do not find the correct entry of bibliographical data. The block format according to ISO 690 is even more compact. The following examples show the layout.

classic three-column form (in Technical Reports):

23. Koller, R.   Engineering design – Basics of design methodology.
                    2nd ed. Berlin, Heidelberg: Springer, 1985.

1st 2nd 3rd building line

space-saving two-column form (in journals and books):

3. *Koller, R.:* Engineering design – Basics of design methodology. 2nd ed.
   Berlin, Heidelberg: Springer, 1985.

1st 2nd building line

Block format according to DIN 1505:

Koller 1985 KOLLER, Richard: *Konstruktionslehre für den Maschinenbau – Grundlagen des methodischen Konstruierens.* 2. Auflage. Berlin, Heidelberg: Springer, 1985.

Riehle/Simmichen 1997 RIEHLE, Manfred ; SIMMICHEN, Elke : *Grundlagen der Werkstofftechnik.* Stuttgart: Deutscher Verlag für Grundstoffindustrie, 1997.

Block format according to ISO 690:

Graham, Sheila. College of one. New York : Viking, 1967.

–. The real F. Scott Fitzgerald thirty-five years later. New York : Grosset & Dunlap, 1976.

Basically the **bibliographical data** shall **allow** the readers of your Technical Report, that they can **find the literature**, which you cited in **libraries**, order it from publishing companies, industrial companies, and institutional bodies **or buy it in a book shop**. It is a matter of fairness, that all required data is provided correctly and completely. You do not need to explicitly specify addresses of publishing companies, industrial companies, and institutional bodies, because their addresses can be found in publically available sources like Yellow Pages and the internet. You can find the address of publishing companies also in the impress of a journal by that publishing company or in the internet.

The list of references has **three blocks of information** which correspond to the columns. **In the first column** there is the **running number** of the source of literature. This column should be right-justified. . The numbering of the list of references can be equal as in the text, e. g. [1], [2], [3] or /1/, /2/, /3/ or (1), (2), (3) etc. This is conform with DIN 1422, part 2, but this is not recommended in ISO 690. ISO 690 recommends the numbering scheme "1., 2., 3. etc."

In the **second column** the **author names** are listed. There are the surnames of the authors with usually abbreviated first names or the long citation. Academic titles of the authors are left out. The publications may have been printed by institutions instead of authors. Then the entries in the second column change, see **Table 3-5**.

The **third column of the list of references** is reserved for the **bibliographical data**. The sequence and structure of these bibliographical data depends on the type of literature and the referencing style.

A number of organizations and institutions have created styles to fit their needs. Publishers also often have their own in-house variations. **Table 3-4** lists some popular referencing styles in various fields of science.

Table 3-4   Popular citation styles in mathematics, natural sciences, engineering and
                  social sciences (each institution have their own variation of typography)

Citation style name	recommended by	Characteristics
Chicago Style	Chicago Manual of Style (CMOS)	used by writers in many fields www.chicagomanualofstyle.org sorting in articles: by order of citation sorting in books: in alphabetical order
Harvard referencing	British Standards Institution, Modern Language Association, American Psychological Association	parenthetical referencing (author-date system) http://www.imperial.ac.uk/Library/pdf/ Harvard_referencing.pdf sorting: in alphabetical order
MHRA Style	Modern Humanities Research Association	note referencing with bibliographic data in footnotes and list of references sorting: by order of citation
ACS style	American Chemical Society	parenthetical referencing (citation numbers) sorting: in alphabetical order
AIP style	American Institute of Physics	parenthetical referencing with literature numbers sorting: alphabetically by author names
AMS style(s)	American Mathematical Society	parenthetical referencing (author's initials and year, e. g. [AB90]), implemented in LaTeX sorting: in alphabetical order
Vancouver system	Council of Science editors, American Society of Mechanical Engineers	citation numbers in square brackets sorting: in alphabetical order
IEEE style	Institution of Electrical and Electronics Engineers	citation numbers in square brackets sorting: by order of citation

If a book has not been written by one author, or a group of authors, but consists of many individual **contributions by different authors**, this is called **host document**. In the second column of the list of references instead of an author's name there is the name of the editor with the note "(ed.)".

If there is no identifiable person as editor of the host document, but an **institution (a corporate body)**, a short name of the institution with the location of the institution's head office in brackets appears as author in the second column. In the third column after the title there is the note "edited by <institution>", that means e. g. "edited by BBC (London)". If several institutions are involved you can solve this as follows: "edited by Siemens AG (Hannover); Nixdorf AG (Paderborn)" or "edited by NDR (Hannover) and University of Hannover".

Table 3-5   Author specification in the list of references

Number and description of authors	Entry in the list of references (examples)

• one author
• two or three authors
• more than three authors
• no author
• standards
• company publications
• institutional bodies
	MILLER, K.
	SMITH, J.; SEBASTIAN, S. and KLING, M.
	MILLER, K. et al. (= Latin: and others)
	Dictionary of librarianship (= title words)
	ISO 4762
	Bosch Rexroth Corp.
	in second column: BBC (London), IEEE
	   (New York), BMBau (Berlin) etc.
	and in third column:
	   ed.: Ministry for regional planning, building and
	   urban development, commission for municipal
	   gardens

Checklists 3-9 and 3-10 show examples and the **data structure** of the bibliographical data for the most common publication types. When specifying authors, editors, and institutions as well as other information **abbreviations** may occur. Here the standards ISO 4 "Documentation – Rules for the abbreviation of title words and titles of publications" and ISO 832 "Documentation Bibliographic references – Abbreviations of typical words" give more information and rules, how words should be abbreviated. **Table 3-6** shows a few examples.

The **typography of the bibliographical data** is partly optional and partly mandatory. It is optional, whether you want to write author's names always in small caps and titles in italic print. But a common system should be used **consistently**. For the punctuation there are **common rules**.

- At the end of the author name(s) and first names there is always a full stop.
- At the end of a title there is always a full stop.
- After the location of the publisher there is a space character and a colon.
- The next item is the publisher, but the word "publisher" is left out for well-known publishers. You may even leave out words from the name, if the publisher is still identifiable. Then you just write "Wiley" instead of "John Wiley & Sons". This item is closed with a comma.
- After the comma you specify the edition and year. To specify the first edition is unusual.
- After the year there is a full stop.

If the information listed above is ambiguous, you may add notes to be more precise. Examples: Cambridge/*UK* or Cambridge/*Mass.*, Available from NTIS : AD 683428.

Table 3-6 Usual abbreviations of terms in bibliographical data of publications

Bibliographical data	Abbreviations according to ISO 832
book	bk.
catalogue, catalog	cat.
collaboration	collab.
collection	coll.
document	doc.
editor, edition	ed.
manuscript	ms.
page	p.
pages	pp.
privately printed	priv. print.
supplement	suppl.
volume	vol.

Checklist 3-9 shows a referencing style of a publisher as an example.

Checklist 3-9 Examples for the bibliographical data of common publication types in Springer format

Journal article (printed)
Smith J, Jones M Jr, Houghton L et al (1999) Future of health insurance. N Engl J Med. 965:325–329
Journal article (online, only by DOI)
Slifka MK, Whitton JL (2000) Clinical implications of dysregulated cytokine production. J Mol Med. doi:10.1007/s001090000086
Journal article (online, no DOI available)
Marshall TG, Marshall FE (2003) New treatments emerge as sarcoidosis yields up its secrets. ClinMed NetPrints.
http://clinmed.netprints.org/cgi/content/full/2003010001v1.
Accessed 24 June 2004

Book (Monograph)
South J, Blass B (2001) The future of modern genomics. Blackwell, London
Book chapter (Contribution to host document)
Brown B, Aaron M (2001) The politics of nature. In: Smith J (ed) The rise of modern genomics. 3rd edn Wiley, New York

No commas between names and initials, no periods after initials or abbreviations.

Here are examples for the data structure of bibliographical data of different publication types according to ISO 690 and ISO 690-2 and example entries in a list of references.

Checklist 3-10 Data structure for bibliographical data of common publication types
according to ISO 690/ISO 690-2 and example entries

Data in the data structure specifications given in italic print may be omitted.

Journal article:
Surname(s) and first name(s) of author(s). Title. *Others involved (e. g. photographer).* In: Title of the journal or publication. series, volume, etc., volume number (year of publication) issue number, pp. (first and last page number of contribution)

McGUIRE, Gerald. *Cyclotron design and efficiency.* Journal on treatment of waste. 112 (1977) 9, pp. 12-20

WEAVER, William. *The collectors : command performances.* Architectural digest. 42 (1985) 12, p. 126-133.

Book (Monograph): Surname(s) and first name(s) of author(s). (Edited by). Title. (series), (volume). *Others involved (e. g. translator, editor).* Location(s) : publisher, edition (if not 1st edition), year of publication. If there is a CD-ROM, DVD, or video cassette as a supplement to the book, you have to add a note at the end of the bibliographical data, e. g. "incl. 1 DVD", *ISBN-number.*

LOMINIADZE, DG. *Cyclotron waves in plasma.* Translated by AN. Dellis; edited by SM. Hamberger. 1st ed. Oxford : Pergamon Press, 1981. 206 p. International series in natural philosophy. Translation of : Ciklotronnye volny v plazme. ISBN 0-08-021680.

Contributions to host documents: Surname(s) and first name(s) of author(s). Title. In: Surname(s) and first name(s) of author(s). (Edited by). Title. (volume). Location(s) : publisher, edition (if not 1st edition), year of publication. (section or chapter number and title, first and last page number of contribution), *ISBN-number.*

Contribution to a scientific series: Surname(s) and first name(s) of author(s). (Edited by). Title. In: Title of the series. (volume). Location(s) : publisher, edition (if not 1st edition), year of publication. *(other data, ISSN number).*

Part of a monograph:
PARKER, TJ. and HASWELL, WD. *A text-book of zoology.* vol. 1, revised by WD. Lang. London : Macmillan, 5th ed., 1930. Section 12, Phylum Mollusca, p. 663-782.

Contribution to a monograph:
WRIGLEY, EA. Parish registers and the historian. In STEEL, DJ. *National index of parish registers.* London: Society of Genealogists, 1968, vol. 1, p.155-167

Publications by companies, institutions, and other corporate bodies:
Short name of the company or institution. Title. Edited by name of the editing company or institution (location of their head office). Publication number or similar data, Location(s) : year of publication.

Standards:
Standard type and number (-part number):year of publication, Title. Location : publisher.
See Chapter 7 of this book.

Intellectual property rights (e. g. patents, trademarks etc.):
Surname(s) and first name(s) of author(s)/inventor(s) or company name. Title. *Others involved.* Note "Intellectual property right:", country code, document number, document type, date of publication. Evtl. patent/trademark holder(s).
CARL ZEISS JENA, VEB. Anordnung zur lichtelektrischen Erfassung der Mitte eines Lichtfeldes. Inventors : W. FEIST, C. WAHNERT, E. FEISTAUER. Intellectual property right: Int. Cl.^{3}: G 02 B 27/14. Schweiz, 608 626. Patent 1979-01-15.
Larsen CE, Trip R, Johnson CR, inventors; Novoste Corporation, assignee. Methods for procedures related to the electrophysiology of the heart. US patent 5,529,067. 1995 Jun 25.
European patent: EP 2013-B1 (1980-08-06)
German patent: DE 27 51 782 1977-11-19

Electronic monograph, data base or computer program: Surname(s) and first name(s) of author(s). Title. Type of medium (database, monograph, computer program, bulletin board, electronic mail), storage of medium, (CD-ROM, online, magnetic tape, disk), *others involved,* edition (if not 1st edition). Location : publisher, year of publication, date of last modification/update or version, date when cited, evtl. also time, *other data as specified below in this checklist like access data, ISBN/ISSN number.*
CARROLL, Lewis. Alice's Adventures in Wonderland [online]. Texinfo. ed. 2.2. Dortmund (Germany): WindSpiel, November 1994 [cited 30 March 1995]. Chapter VII. A Mad Tea-Party. Available from World Wide Web: <http://www.germany.eu.net/books/carroll/alice_10.html#SEC13>.

Contribution to a source of literature from the internet: Surname(s) and first name(s) of author(s). Title. In: series, evtl. volume, issue number, first and last page number of contribution. DOI number or information regarding the last update and/or version, URL/URI or starting URL/URI and description of clicks you have to perform to go to the desired page, date when cited, evtl. also time.

Electronic mailing lists, discussion forums: Title. Type of medium, Location : publisher or owner of the forum, date when published, date of the citation, *other data,* access data
Parker, Elliot. Re: Citing Electronic Journals. In Pacs-L (Public Access Computer Systems Forum) [online]. Houston (Tex.): University of Huston Libraries, 24 November 1989; 13:29:35 CST [cited 1 January 1995; 16:15 EST]. Available from Internet: <telnet://brsuser@a.cni.org>.

Additional data for personal messages and e-mails: Title of the contribution or e-mail. Type of medium or location. *Others involved/list of addressees/event,* date of e-mail delivery or event.

☞ *General rule in ISO 690: First names can be abbreviated by initial letters. "Miller, William Thomas" will then be listed as "Miller, WT.".* To distinguish the elements in the bibliographical data you should apply **a consistent system of typography and punctuation**, e. g. that every element is finished with a period. **It does not matter, which system you apply.** *The main point is to use it consistently.*

☞ *Other recommendations in ISO 690: A subtitle is distinguished from the title with a colon. Several publisher's locations are separated with a semicolon. The year or date of publication is specified as in ISO 2014 or as in the source of reference. Country, province and state names, which specify the names of cities more precisely can be abbreviated according to ISO 3166. Title words can be abbreviated according to ISO 4.*

☞ *Not all listed data is mandatory. If in doubt, please refer to the standard. One source for both parts is:* http://openpdf.com/ebook/iso-690-1-pdf.html.

Now we want to give you some hints and information regarding **special cases**.

Special cases

If you cite from a **brochure or manufacturer document** and bind this document together with your Technical Report, please add a note in your list of references like "..., see Appendix C Other sources".

If **sources of information are not accessible to the public**, this must be clearly specified. You can add notes like "**(in print)**" or "**oral statement/comment**" or "**citation from an e-mail** written on 07.Jun.2005", "**NDR radio transmission** ‚Auf ein Wort' on 12.Mar.2006" etc. In case of oral statements/comments it should be stated when exactly at which meeting in front of which audience the statement/comment was given (during a presentation, in a certain TV discussion etc.).

Additional information, which is useful for finding or buying literature (like ISBN for books, ISSN for journals, DOI for electronic documents, Library of Congress Control Number LCCN for documents registered by the American National Library, ordering number of a book etc.) can be added at the end of the bibliographical data.

If bibliographical **data for articles in journals cannot be provided in the structure listed above,** you should deliver similar information. Sometimes all required bibliographical data is printed on each page of the journal. But quite frequently on the single pages of the journal the journal title is listed (often abbreviated) and the issue number, but the volume number by the publisher is missing. However, volume numbers are very important for finding journals in libraries, because at the end of the year the journals are bound as books. If there is **only one volume per year**, the libraries use the same volume number as the publishers and the volumes are identified by their **volume number**, e. g. "vol. 54". If there are **several volumes** per year and the page numbering runs through from the first issue to the last, the volumes are identified by their issue numbers or by **page numbers**. In this case the libraries use their own volume numbers. The bibliographical data of a single (fictive) article from volume 37 of the journal could then look like this:

33. LIEHR, J., *Thermochemical Gasification of wood as useful removal of waste wood*. In: Journal on treatment of waste. 37 (1985), pp. 824-836

Those who are looking for this article can look up the volume number by the publisher in the data base in the library. Using the data base is mandatory, because each library decides for themselves, into how many volumes the issues of one year are split.

If the issue number, year of publication and volume number are not printed onto the individual pages of the issues, the missing data may be found in the impress. The **impress** is often in the very front or back of an issue, in or near the table of contents. The journal volume numbers of the libraries can be found in the catalogues of the libraries.

Another problem is, that journals **sometimes have no issue numbers**, but other issue labels. Then you should add these issue labels accordingly to the list of references, here Aug./Sept.

Example: 72 (1990), Aug./Sept., pp. 115-117

If a piece of **information is published in a data network**, it can be used by any other user in the network (provided the right access permissions are set). The users can save an electronic copy on your harddisk drive or a storage device, use and modify the piece of information as a whole or parts of it in any form.

These users could e. g. offend against copyright laws by deleting the author's name, using the information for commercial purposes without paying license fees, passing on the information to other users in the data network without prior asking for permission etc., and the temptation to do so is quite large. The rules in **Checklist 3-11** and the information in 3.5.7 "Copyright and copyright laws" help you to behave correctly.

Checklist 3-11 Dealing with information from data networks

1. If the author or publisher of a piece of information formulates some **limiting conditions**, you should be fair and follow these conditions.
2. **Citations from e-mails** are treated as citations from physical letters or oral statements/comments (name the author and specify type and date of publication, do not manipulate the sense of the source text).

3. **Documents from the internet** can be cited, as other sources of literature, literally or analogously. In both cases the bibliographical data must contain surname and first name(s) of the author, title. Either DOI number or date of last update or version, URL/URI or starting URL/URI and list of clicks you have to make to reach the desired page and the note "Accessed on: <date, evtl. also time>".

   By the way: To limit the distribution of illegal contents, everybody is responsible for the information, he/she makes accessible via links.

4. If you wish to use information from the internet, you should think about whether you want to cite or link the information. **If you want to cite it, download the information completely with all image files to your hard disk drive or another storage device.** Note the URL/URI or the click sequence and the date, when you downloaded the information, because the operators of the homepage can change or delete the contents at any time. If you just want to link the information, the URL/URI and date of access is sufficient.

5. Computer programs can also be copied via data networks. You should obey the conditions for freeware/shareware (only complete, non-commercial transfer to third parties is allowed, license fees should be paid).

These questions how to cite information from data networks correctly are relevant for any author of Technical Reports, who has access to such data networks (mainly the internet). If you have questions regarding complex texts or mathematical deductions you can ask for assistance in the web. If there are problems with the material collection, colleagues in the web may help. In principle, any information can be transferred via data networks. You as the author of a Technical Report should always specify "all used help and sources" truthfully to be on the safe side.

After the bibliographical data has been collected, we want to give you some hints regarding **the typographic design of the list of references** now. **Lists of references** are written with **single spacing**. However, if your supervisor always expects that a thesis must have 60 pages and you have only 55 pages, you can vary the spacing and deliver the required page number. Here is some flexibility. Since the transfer of the general rules to the individual sources of literature is hard for many authors, **an example list of references** shall facilitate this transfer, **Table 3-7**. Then we will give you a list with publication types and publication numbers in the example list of references, **Table 3-8**.

Table 3-7   Example of a three-column list of references <to be continued>

---

9 List of references

1. BOSCH   Global Responsibility – Environmental Report 2003/2004.
                         www.bosch.com/content/language1/downloads/UWB_en.pdf
                         Accessed on: 04.Jun.2006

2. YATES, JG.   Fundamentals of Fluidised Bed Chemical Processes.
                         Butterwoths Monographs in Chemical Engineering.
                         London : Butterworths, 1983

3. KUNII, D. and   Fluidisation Engineering. New York : Wiley, 1969
    LEVENSPIEL, O.

4. DAVIDSON, JF.,   Fluidisation. London : Academic Press, 2nd ed. 1985
    CLIFT, R. and
    HARRISON, D.

5. ORMOZ, Z.,   Studies on granulation in fluidised bed V. – Study on the
    CSUKAS, B.   particle size distribution of granulates.
    and PATAKI, K.   Hung J of Ind Chem Veszprem, Vol. 3 (1975) pp. 193-215

6. WALDIE, B.,   Kinetics and mechanisms of growth in batch and continous
    WILKINSON, D.   fluidised bed granulation. Chem Eng Sci, Vol 42 (1987),
    and ZACHRA, L.   No. 4, pp. 653-655

7. LAPPLE, CE.,   Atomization – A survey and critique of the literature.
    HENRY, JP.   Stanford research institute, Menlo, California, April 1976,
    and BLAKE, DE.   available as microfiche AD 821 314 from British Library

8. HAUSER, EA. et al. The application of the high-speed motion picture camera to
                         the research on the surface tension of liquids. J Phys Chem,
                         Vol. 40 (1936), pp. 973-988

9. KDG Flowmeters   Calibration chart for variable area glass tube size 14 and 14X,
                         type E, F and G from KDG flowmeters, A division of KDG
                         Instruments Ltd., Rotameter Works, 330 Puley Way,
                         Croydon, Surrey, England, 1989

10. HERING, H.   Powder Granule Generation in a Fluidised Bed
                         (Pulvergranulaterzeugung in Wirbelschichten).
                         Diploma thesis, Heriot-Watt-University, Department for
                         Chemical and Process Engineering, Edingburgh/GB : 1989

11. SKF   SKF Hauptkatalog (main catalogue). Catalogue 4000/IV T,
                         Schweinfurt, 1994

12. Wikipedia   Toluol. www.wikipedia.org/wiki/toluol,
                         last update: 30.Apr.2006, accessed on: 04.Jun.2006.

Table 3-7   Example of a three-column list of references <continued>

13. Statistics Canada	Communications equipment manufacturers. Manufacturing and Primary Industries Division, Statistics Canada. Preliminary Edition.1970- Ottawa : Statistics Canada, 1971- . Annual census of manufacturers. Text in English and French. ISSN 0700-0758.
14. WEAVER, W.	The collectors : command performances. Photography by RE. BRIGHT. Architectural Digest. December 1985, Vol 42, no. 12, pp. 126-133
15. HERING, L. and HERING, H.	How to write Technical Reports. Seminar documentation. 1998
16. BÜRGI, F.	Scope and usage of multimedia in training and education. In: MELEZINEK, A. (ed.): The engineer in the joined Europe – Reflections and perspectives. 20 years IGIP. Presentations of the 21st International Symposium "Engineering Paedagogics '92". Leuchtturm-Schriftenreihe Vol. 30 (1992) Alsbach/Bergstraße : Leuchtturm-Verlag, pp. 221-226
17. BS 188	Specifications for viscometers. 1977

Next comes an analysis of the three-column list of references above by type of the cited literature, **Table 3-8**. If you collect the list of references for your Technical Report and you are uncertain how to enter the bibliographical data, you can look up the type of publication and the appertaining literature numbers in **Table 3-8**. Then you go to the example list of references, **Table 3-7**, and enter the data. You have to decide, whether your publication is a book as monograph, a book from a series, an article in a journal or an article as a contribution in conference or symposium proceedings. The other publication types are easier to distinguish. If the publication you want to cite does not exactly match the publication types in **Table 3-8**, please go back to **Checklists 3-9 and 3-10**, collect the bibliographical data and enter it accordingly into your list of references.

The document part **"Literature" or "References" at the end of an article in a journal** contains the same bibliographical data, but the layout is much more space-saving, **Table 3-9**. It depends on you, whether you want to apply the systematic and layout of the „space-saving form" (two columns) or the block format according to DIN 1505 with a repetition of the citation or the block format according to ISO 690. It is also up to you, to type author's names in capital letters or small caps and titles in italic print.

Basically, the rules of university institutes, companies, and other customers, i. e. the **"rules of the house" must be followed**. If you write a book or an article, the rules of the **publishing company** apply.

Table 3-8   Type of publication in the list of references in three-column layout

Type of publication	Entries in the example list of references
book, as monograph:	2, 3, 4
book, in a series:	13
contribution to a series (here a photography, similar for a chapter, section etc.):	14
research report available on microfiche:	7
article in a journal:	5, 6, 8
article as contribution on a conference or symposium:	16
seminar or lecture documentation:	15
standard:	17
company report:	1
dissertation, habilitation, study report, master, diploma, bachelor thesis:	10
literature from the internet:	1, 12
data sheet:	9
manufacturer catalogue:	11

☞ *The correct setup of a list of references lasts much longer than estimated. Often missing bibliographical data must be collected. The strict rules regarding layout and sequence of information reduce the typing speed drastically. Therefore you have to plan very large time reserves for writing the list of references. In case of doubt, the completeness of the bibliographical data is more important than following all the layout and sequence rules.*

Table 3-9   Example of a space-saving list of references

5 References

1. BOSCH, *Global Responsibility – Environmental Report 2003/2004.*
    www.bosch.com/content/language1/downloads/UWB_en.pdf
    Accessed on: 04.Jun.2006

2. YATES, JG. *Fundamentals of Fluidised Bed Chemical Processes.*
    Butterworths Monographs in Chemical Engineering. London : Butterworths, 1983

3. KUNII, D. and LEVENSPIEL, O. *Fluidisation Engineering.* New York : Wiley, 1969

4. DAVISON, JF., CLIFT, R. and HARRISON, D. *Fluidisation.* London : Academic Press,
    2nd ed. 1985

5. ORMOZ, Z., CSUKAS, B. and PATAKI, K. *Studies on granulation in fluidised bed V. –
    Study on the particle size distribution of granulates.* Hung J of Ind Chem
    Veszprem, Vol. 3 (1975), pp. 193-215

In the next section we shortly present you some rules from ISO 690 to work with documents written in foreign languages.

3.5.6 Working with documents written in foreign languages

Bibliographical data in languages that do not use Roman letters shall be transliterated/romanized in accordance with the appropriate standard: "Medicinska akademija" or "Медицинска академия (Medicinska akademija)". The following ISO standards apply:

- ISO 9, Documentation – Transliteration of Slavic Cyrillic characters into Latin characters
- ISO 233, Documentation – Transliteration of Arabic characters into Latin characters
- ISO 259, Documentation – Transliteration of Hebrew characters into Latin characters
- ISO 843/R, Documentation – Transliteration of Greek characters into Latin characters
- ISO 7098, Documentation – Romanization of Chinese

If a title in the original language (e. g. the title of a book, an article or a journal) cannot be understood by the audience without doubt it should be translated and the title in the original language added in brackets behind the English data.

3.5.7 Copyright and copyright laws

On international level the copyright law is very much influenced by the Berne Convention for the Protection of Literary and Artistic Works, which is published at http://www.wipo.int/export/sites/www/treaties/en/documents/pdf/berne.pdf.

The Berne Convention is binding under international law since 1886. It has undergone several **revisions. Therefore since 1908 people call it Revised Berne Convention (RBC)**. Article 5.1 states, that every country which has signed the treaty, accepts the copyright of publications created by inhabitants from other treaty partners in the same way as it accepts the copyright of publications created by the own inhabitants. This equal ranking of foreign authors and domestic authors makes it unnecessary to investigate in foreign copyright law details, because the protection of authors' rights comes into effect automatically without any registration or copyright annotation. The RBC guarantees the copyright protection for at least fifty years after the author has died. The EC has elongated this period to 70 years in 1993 to harmonize the copyright regulations in Europe. The United States followed this example with the Sonny Bono Copyright Term Extension Act in 1998.

Since 1967 the Berne Convention is managed by the World Intellectual Property Organization (WIPO). The World Trade Organization (WTO) has passed the Agreement on Trade-Related Aspects of Intellectual Property Rights (TRIPS) at the end of the Uruguay Round of the General Agreement on Tariffs and Trade (GATT) in 1994 for its member countries. Since most countries in the world are WTO members, even countries, which did not sign the Berne Convention, are obliged to accept many of its conditions and regulations.

The copyright laws in the English-speaking world regulate, that an author can publish his independent works or let them print by a publisher. The author has to get royalties, if someone else publishes his work. This legal right cannot be inherited. And the copyright laws in the English-speaking countries are less strict than in Europe (namely in Germany) – in other details as well. Therefore we want to describe the German legal situation and recommend that you arrange yourself with that situation to be always on the safe side.

In Germany the copyright law Urheberrechtsgesetz (UrhG) is *"the independent right of authors and creators of cultural and scientific works on any usage and liquidation of their works and the protection against improper use of their works"*.

This law protects works created in the fields science, literature, music, dancing, fine arts, architecture, arts and crafts, photography and cinema. In the context of science and technology the copyright law mainly protects books and contributions to books, articles in newspapers and journals, figures, computer programs, publications in the internet, video films etc. In opposition to the fact that technical ideas are a part of our modern culture as well, technical ideas are *no* cultural works in the sense of the copyright law. They can only be protected as patent, trademark or design patent. Software cannot be patented. To protect software, you can apply for a trademark and a word mark. Thus the copyright law does not cover technical ideas. The following information is derived from several books and internet sources and cover the latest changes of the copyright law in 2003 and 2007, i. e. the legal situation which is valid since 1.Jan.2008. The following sources have mainly been used:

- NORDEMANN, W.; VINCK, K.; HERTIN, P. W.: Urheberrecht – Kommentar zum Urheberrechtsgesetz und zum Urheberrechtswahrnehmungsgesetz.
- HUBMANN, H.: Urheber- und Verlagsrecht.
- ILZHÖFER, V.: Patent-, Marken- und Urheberrecht – Leitfaden für Unterricht und Praxis.
- Harsh criticism on the last changes of the law, esp. regarding § 52a and § 53: http://www.52a.de

In principle the works protected by the copyright law may only be looked at or read. Under certain conditions they may also be copied, typed, scanned, redrawn, saved on a data storage device and cited. However, they may not be copied, published, transmitted or performed as a whole without permission of the author. The copyright law wants to give the author complete control over all possible uses of his work.

To permit scientists and persons engaged in the cultural sector that they can work without too narrow limits there are some exceptions from this superordinated right. In the following the exceptions that concern teaching, science and technology are described in more detail.

The most important exception is the right for anybody, to make **copies for one's own private and other personal use** for the following purposes:

- for one's own scientific research (no commercial use),
- for adding the copies to one's own archive (either paper copy + digital copy or analogue copy, the archive must serve public interest and do not have commercial interest)
- to become informed about current news (radio and TV transmissions only),
- and for other personal use, if only small parts of the work or single contributions are copied or if the work is sold out since at least two years.

Copies of small parts of a larger printed work, e. g. of a book or of individual articles from a newspaper/journal may also be made for teaching, for examinations and for **non-commercial use** in the fields training and education as well as vocational training. To be sure you should mark the copies with **"for teaching purposes"**. These copies, which you made for your own use or for teaching purposes may not be made available to the public or performed in the public, § 53 UrhG.

This paragraph of the copyright law was revised 2003 and adopted to the digital world. Therefore, it is **allowed to make a digital copy** and use it in the sense described above. You may publish a digital copy in an intranet or send it via e-mail, if this is important for the exchange within a group of scientists or a closed group of pupils or students. **Publishing** a digital copy **via internet** is equivalent to a public performance or transmission and thus **not allowed**. An essentially complete graphical copy of sheet music or **an essentially complete copy of books or journals and computer programs** may **only** be made **with the permission of the author**!

☞ *It belongs to scientific working to use other works, i. e. to co-use the findings and intellectual creations of others together with one's own findings and creations. This co-use can be pure reading to get an inspiration or copying **small portions of the other work** and citing the source. As a rule of thumb, ≤ 5 % of the text or the images of the original source are "a small portion". If a work produced by someone else is cited in major parts and only slightly modified, the agreement and permission of the other author is required, and if the author wants to use the work by the other author completely or nearly completely, he has to pay the other author appropriate royalties (according to § 54 UrhG).*

☞ *Citing is allowed without charge. A **literature citation is copying another work** as a whole or in parts without changing the sense and using it in an independent new (scientific) work **and correctly specifying the source**. Literature citations may only be made in independent works which again are protected by copyright law. This new work must be a personal intellectual creation of the* *citing author, i. e. his own achievement. The literature citation is **only allowed,** if in the new work it plays **a very subordinate role**.*

According to the definition of a literature citation as an unaltered copy from the source you may not change the meaning, neither in literal nor in analogous citations. Also translations to other languages are only allowed, if the meaning of the source persists. Moreover, the literature citation is only allowed, if it

- supports the author's own position,
- is an example for an author's own statement,
- makes clear analogue opinions,
- explains the own descriptions or
- it gives reasons or deepens the own statements.

As an addition to the statements in the books named above we would like to add the following: It is also allowed to cite opposite opinions, differing statements and current discussions in the scientific community. The author then argues these cited opinions and gives reasons for his own opinion or his selection of dates and facts.

☞ *It is not allowed to use a literature citation just as rounding off, addition or completion. The following check might help you as rule of thumb: If there is no relationship between the literature citation and the new work, it is nearly always not allowed to use the literature citation. You are on the safe side, if you go into the cited information and respond to it by working out the relationship to your statements and findings.*

Another **basic condition** for the permissibility of a literature citation is a **correct and complete citation** in the text. If leave out information or change the meaning or message of the cited work, this must be specified.

How to mark the citations (notes of reference) in the text and how to put up a correct list of references is described in sections 3.5.4 and 3.5.5.

Beside the information above the following regulations are listed in the copyright law:

- The copyright ends seventy years after the author's death or
- in case of several authors seventy years after the last author's death.
- After the author has died, the copyright is transferred to his inheritors.

In the practical application of the legal regulations described above there are often small **deviations** and seldom larger **offenses**. However, if in a bulky script of an academic lecture about 50 images are copied without any citation (note of reference) and the work where the images have been copied from is not listed at all in the list of references, this is either a severe carelessness or a deliberate offense against the copyright law.

The **copyright law was further changed** to integrate the EU regulation 2001/29/EG into German law. Since 1.Jan.2008 The changes are part of the UrhG. **Compared with the situation in 2007 there are strong limitations now**, because the change of the UrhG in 2003 resulted in a reduction of sales of school books and text books by up to 80%. Now the following rules apply:

- Private copies on paper or on computers in non-editable file formats remain allowed.
- Digital copies of works, which are equipped by industry or publishers with digital rights management and copy protection (e. g. films, pieces of music, video games), may not be copied. The copy protection may not be broken.
- The copy of an original work which has obviously been made illegitimate and the copy of an original work which is obviously provided for download unlawfully (e. g. on a file sharing platform/network) are accusable.
- There was a flat-rate on copiers, printers, and storage media as a compensation for the still allowed private copies. The height of the flat-rate was legally limited. Now the author's collecting societies and the associations of the manufacturers of copying devices and storage media shall agree on the royalties themselves. The legal limitation of the royalties does no longer apply. Maybe mobile phones, digital cameras, printers, copiers, computers and their accessories will become more expensive.
- Libraries, museums and archives may display copyright-protected digital (i. e. editable!) copies of works at special computers (no commercial use!).
- Schools may put small parts of works protected by copyright law on school servers, if it is not a part of a work specially made for educational purposes and the class or group of pupils and the learning unit are predefined (no saving of documents on stock). The pupils and students may only read the documents during their classes (not from home, in spare time etc., copy protection must be provided). Usage of school books is never allowed without permission of the publisher (not even if only one image shall be used).
- Electronic materials of libraries and information providers may only be sent to users as graphic files. Sending digital copies to external users is only allowed, if the publisher has no obvious own online offer (pay-per-view) for appropriate conditions.

Many parties want to influence the copyright law. The interests of manufacturers of copying and storage devices, publishers, authors, translators, readers as well as libraries, museums, archives and information providers differ very much. A fair agreement will be hard to find.

3.6 The text of the Technical Report

In Technical Reports (written or as a presentation) the technical terminology is written or spoken, taking the target group(s) into consideration. This technical terminology does not differ from the general language as much as it is often the case e. g. in psychological and sociological, medical and legal texts.

The following sections provide you with hints, how to write good style in general and in Technical Reports, which peculiarities occur in the context of formulas and computations, and how you can improve the understandability of your texts.

3.6.1 Good writing style in general texts

In schools as well as in universities reports, protocols and presentations are written with computers. This implies, that basic knowledge of structuring texts, typography, syntax, style etc. are already taught there. Yet, in reality this knowledge is in most cases improvable. Due to the differing prior knowledge, in this section only those general rules for good writing style are shortly listed, which are useful for all sorts of non-fictional texts. Please remember, that a Technical Report must be written clearly, understandably and oriented towards the target group(s). **Checklist 3-12** shows general rules for a better understandability of texts, which are also valid for other non-fictional texts than *Technical Reports*.

Checklist 3-12 General rules for a better understandability of texts

- Formulate short sentences.
- If possible, use only one main clause with one or two subordinate clauses or two main clauses connected with a comma or semicolon.
- Do not use too many foreign words.
- Explain foreign words and unknown abbreviations when they occur the first time.
- Avoid too many abbreviations.
- Use introductory and connecting sentences, to guide your readers with words. In these sentences you can:
  - refer to the structure or table of contents,
  - conclude the facts described so far,
  - build a connection to the next document part or
  - introduce a new document part.
- Use descriptive formulations, write vividly and engagingly! Analogies, metaphors and comparisons create associations in readers' minds. So readers can recognize similarities with and deviations from their existing knowledge more easily.

One part of these general rules which improve the understandability of texts is the "*iS*", already introduced in 3.1.3 "Text with figures, tables, and literature citations". "*iS*" stands for "introductory sentence". You should avoid that after a document part heading there are figures, tables, or bullet lists directly following the heading without a connecting or introductory text. It is much better, if you use an "*iS*" at the beginning (and end) of each document part. Such introductory and connecting sentences tie in with the **prior knowledge** of your readers and structure the written information. All details are sorted under consideration of the

"backbone". The readers are not left alone, but "guided with words". The frequent use of introductory and connecting sentences is a prerequisite that the **readers understand** your text **without questions** in the sense you want them to. Referring to the rule **"Use descriptive formulations ..."** here are three examples:

- **Analogy:** the orbit of an electron flying around the atom's nucleus <u>is similar with</u> the orbit of a planet travelling around its central star
- **Metaphor:** (= imaginary expression in the figurative sense): a figure of the planet earth often <u>stands for</u> the internet
- **Comparison:** an electron in highly charged condition <u>is like</u> a tense spring.

If you are in doubt whether a formulation is understandable and corresponds to good writing style, you may wish to use Microsoft Word's **grammar checker**. Please refer to Special – Options – Spelling and grammar – Preferences. Activate an option or preference or grammar rule. Word will show you (possible) offenses against this rule by green wavy lies. If an option name is ambiguous to you, mark the option and click on Explain.

But now we want to deal with the rules which should be applied especially for Technical Reports.

3.6.2 Good writing style in Technical Reports

The style of the Technical Report shall follow the general rules for good writing style above. But there are also some rules, which are only valid for Technical Reports and which must be obeyed, too. Above all other rules there is the following basic principle:

☞ *Clarity and unambiguousness have always preference compared with writing style rules!*

The Technical Report shall be understandable without questions for readers, who have technical knowledge but no detailed knowledge of the current project. Since authors of a Technical Report have often worked on their project for weeks or months, they can hardly imagine how much (or better: how little) a normal reader of the report can know about the project at all, when they start to write the report in the last phase of the project. Therefore in Technical Reports **too much detail knowledge is supposed very often**, which the addressees do not have. This overstrains the readers of the report quite frequently, and their motivation to read the report is negatively influenced.

In addition, after frequent reading of own texts authors tend to become blinded by routine against their own formulations. Therefore, during the end check of the Technical Report you should show it a friend or colleague, so that he/she can proof whether the report is understandable for people who have not been involved in the project.

The Technical Report is usually written **impersonally**, i. e. **passive sentences** are used quite frequently and personal pronouns like **"I, we, my, our, you, etc." are avoided.** However, in a summary or critical appreciation it is OK, to speak of "we" or "our", if the own working group or department is meant.

It is **traditionally** like that. Most technicians got used to the impersonal way of writing during their education and professional practice. The customers will probably prefer impersonal writing as well, because they are used to it, too. For non-technicians passive sentences and impersonal writing seem clumsy, boring, and monotonous. Therefore many books about good writing style recommend to write in active voice.

Passive sentences also comprise the **danger, that the reader is uncertain who does something.** For example, in a description of a machine or plant it may be uncertain, whether the operator or an automatic mechanism of the machine or plant executes an action. If this may occur, you should use an addition like "by the operating staff" (man) or "by the revolver control" (machine).

You have to decide carefully for your Technical Report, whether, how much and where you want to use active sentences instead of the usual passive. You are "on the safe side", if you avoid personal pronouns and use the passive voice instead. Here an example of formulating the same fact once in active and once in passive voice:

Active: "... **we** have evaluated the following alternatives ..."
Passive: "... the following alternatives have been evaluated ..."
                          we is <u>bad writing style</u> in Technical Reports!

The **tense** is present tense. Past tense is only used, if a previously used part, measuring procedure or similar is described.

The **naming of technical appliances, assemblies, parts and procedures** (e. g. in part lists and design descriptions) shall be **according to the function of the appliance, assembly, part, or procedure**, i. e. not handle but shutting handle, not gear but input gear, not plate but retaining plate, not angle but stiffening angle, not frame but carrying frame, not variant 1 but electric-mechanical solution etc. The general principle for naming parts is:

☞ *Parts are always named according to their function.*

This holds true in all areas of technology, e. g. also in civil engineering and electrical engineering. Now some examples as an illustration.

Company or working group internal names are probably unknown to the readers. Therefore, select neutral names.

If an assembly drawing and part list or a sketch of the described appliance belong to the Technical Report, it makes sense to add the **position number in brackets after the part name**. That is to say for example: "The fixation of the cover plate (23) was accomplished by groove pins to save costs." Here "23" is the position number and "cover plate" the name from the part list. This correlation of number and name in report and drawing as well as part list helps the reader, to find his way in the different documents.

Quite frequently you can read the following or similar formulations in Technical Reports. "The design has a high mechanical strength and a very good wear resistance." In this case it is better, to substitute the general term "design" with the **actual name of the complete (sub)assembly or plant**. An application of this rule to the example above could be: "The oil mill has ... ."

Often a **value range** is specified with two physical values. Then it is wrong to write: measure and measuring unit, extension sign (– or to) and again measure and measuring unit. correct is to write the dimension only once, after the two measures. An example for illustration (text from the description of a mold for injection molding):

"The main problem is the effective insulation of the hot mold (ca. 160 °C̶ to 200 °C) from the relatively cool distributor duct area (ca. 80 °C̶ to 110 °C)." As a test you can read the written text loudly. Then you will find out fast, if there are unnecessary details in the text.

To finish this section here is a further style rule for Technical Reports. Since the Technical Report is addressed to technicians, who normally approach all problems in a rational way, there should be **no emotional and colloquial formulations** in the Technical Report. Thus, sentences like "After the ergonomic analysis of his workplace the employee can continue to work joyfully." or "The programmed software ran really cool." should better be avoided.

3.6.3 Formulas and computations

Formulas appear in Technical Reports mainly, if computations are developed. These computations can e. g. occur in the fields mathematics, physics and chemistry, but also in core areas of technology like civil engineering, mechanical engineering, electrical engineering, and informatics.

The formulas are often closely interwoven with the text, because the text refers to individual physical values, makes a statement on the formula etc. Text and formulas are often an inseparable unit. Therefore we have reserved "only" a section within the subchapter 3.6 "The text in the Technical Report" for formulas and not an own subchapter like for tables, figures, and literature citations.

In the technical education pupils and students are confronted with teachers and professors using different formula signs for the same physical values in different subjects or lectures. This creates unnecessary disturbances and irritations. Here the standards try to help. Those who would like to go deeper into the fields formula symbols, formula syntax and formula layout can e. g. consult ISO 80000, Quantities and units with 15 parts covering the set of mathematical signs and symbols, the SI and the symbols for chemical elements:

ISO 80000-1   General
ISO 80000-2   Mathematical signs and symbols to be used in the natural
               sciences and technology
ISO 80000-3   Space and time
ISO 80000-4   Mechanics
ISO 80000-5   Thermodynamics
IEC 80000-6   Electromagnetism
ISO 80000-7   Light
ISO 80000-8   Acoustics
ISO 80000-9   Physical chemistry and molecular physics
ISO 80000-10   Atomic and nuclear physics
ISO 80000-11   Characteristic numbers
ISO 80000-12   Solid state physics
IEC 80000-13   Information science and technology
IEC 80000-14   Telebiometrics related to human physiology
IEC 80000-15   Telebiometrics related to telehealth and
               world-wide telemedicines

At first the term formula shall be defined: A formula can consist of formula symbols, physical values without dimension (constants), physical values with dimension (measures and measuring units, between them is exactly one space character) and mathematical or other operators.

formula		
formula symbols	**physical values**	**mathematical and other operators**
vectors: **a, b, c**, ..., **x, y, z** or without bold print with a right arrow above the vector coordinates of vectors: a^1, a^2, a^3, ... scalars: $a, b, c, ..., x, y, z$ other formula symbols: matrices, column vectors, determinants etc. are displayed accordingly	examples: length l in m time t in s velocity v in m/s mass m in kg force F in N pressure p in Pa power P in W concentration c in mmol/l quantity n in mol molar mass M in kg/mol	$+, -, *$ or $\cdot, : /$ or solidus $\pm = < \leq > \geq \sim \approx$ $<< >> \hat{=}$ $\times \perp \parallel \angle \nabla \cap \cup$ $\supset \supseteq \not\subset \subset \subseteq \in \notin$ etc. besides: integral symbol $\int$ root symbol $\sqrt{x}$ exponents 23^6 indices L_{max} and diverse brackets

For these formula symbols the information relevant for Technical Reports is now described. If a text contains only a small number of equations, the equations are emphasized by blank lines before and after the formula and evtl. by an indentation of about 2 cm on DIN A4 paper. If a text contains many equations, the equations are additionally numbered with **equation numbers** like "(16)" or "(3-16)" at the right margin of the printing area, so that you can better refer from the text to the equation. If a formula spreads across more than one line, the equation number appears in the **last line** of this formula.

☞ *If you want to write **formulas in HTML documents**, you can save typing effort, if you visit the URL/URI www.mathe-online.at/formeln.*

Comfortable word processors have a **formula editor**. If you want to type only a few simple formulas, you can use the function Insert – Field – Formula and evtl. the options as well as Insert – Field – Expression in Word.

If you enter **formulas in normal text mode**, you should use the following settings: formula symbols italic, indices and exponents standard and superscript or subscript (= 2 pt smaller than the basic font size); minus as operator for computations: dash (Alt+0150), minus as algebraic sign: normal hyphen, multiplication dot: · (Alt+0183). Example:

$$Q_{up} = c_w \cdot m_w \cdot (T_M - T_W^2) + c_k \cdot (T_M - T_W^2) \tag{16-13}$$

More complex formulas can be typed in easier with a formula editor.

$$\Delta S = C_n \cdot m_n \cdot \ln\left(\frac{T_E}{T_{h,A}}\right) + c_t \cdot m_t \cdot \left(\frac{T_E}{T_{h,A}}\right) \tag{16-14}$$

☞ *Tricks for the formula editor:*
- *Enter space characters with Ctrl + space bar.*
- *If ascenders and descenders are not printed completely (e. g. the upper line of the root symbol), select a slightly larger line spacing!*
- *The proportional symbol ~ can be typed in normal text and formula editor as AltGr-+.*
- *The formatting templates Matrix/Vector and Text result in non-italic text (e. g. for measuring units).*

If you have to write many formulas, you can also use the layout program "TEX" or the appertaining macro package "L^ATEX". It creates a very smooth and pretty formula layout. Some more image examples can be found in the Wikipedia article "Formula". Many publishing companies demand this format or offer at least layout templates for this format. The program is available as public domain software and widely spread in universities. An option to use the advantages of L^ATEX without having to install it on your computer is a Knoppix- or Kanopix-CD. You start your computer with the CD in the drive and the computer starts as Linux-PC. Depending on the distribution you can then use L^AT_EX directly, without having to install it. Ask someone in the next Linux club (or a "knowing" friend).

Formulas appear seldom as an individual formula, but mostly in **more bulky computations with several formulas**. The text should explain the formula symbols, describe formula transformations, pick up the results of the computations etc. It is desirable to name and explain the physical values, which appear in a formula, in the text shortly before or after the formula.

Here is an example of fracture mechanics computations:

IRWIN [28] developed another approach than GRIFFITH [23] to forecast the brittle fracture probability of ductile materials. IRWIN considered the elastic stress field ahead of the crack, which is characterized by the stress intensity factor K. With the stress σ and the crack length a it is:

$$K = \sigma \cdot \sqrt{\pi \cdot a} \cdot f(a/W) \qquad (2\text{-}15)$$

The factor $f(a/W)$ is dimensionless. It depends on the geometry of the specimen and the crack. Values for $f(a/W)$ are tabulated in standards, e. g. in BS 5447. Crack extension occurs, if the factor $\sqrt{\pi \cdot a}$ reaches a critical value, the so-called critical stress intensity factor

$$K_c = \sigma_f \cdot \sqrt{\pi \cdot a} \cdot f(a/W) = \sigma_f \cdot \sqrt{a} \cdot Y \qquad (2\text{-}16)$$

where σ_f is the fracture stress. The factor Y is introduced for simplification. It is computed as

$$Y = \sqrt{\pi} \cdot f(a/W) \qquad (2\text{-}17)$$

The critical stress intensity factor K_c is obtained out of the experimental data very easily, since K_c^2 is given by the slope of a straight line of σ_f^2 against $1/a$, see equation 2-16.

The stress intensity factor refers to a specific load state. It is expressed via indices. I stands for normal stress, II for shear stress, perpendicular to the crack tip and III for shear stress parallel with the crack tip. The critical stress intensity factor for normal stress K_{Ic} is also called fracture toughness. If the plastically deformed zone at the crack tip is small compared with the geometry of the specimen, this value is independent of the specimen shape and size and therefore a material constant.

In such computations the formulas, the computed values and the text are connected with each other by **signal words** like *"With* <physical value> *and* < physical value > *the result is* <formula>*."* and *"Taking into consideration* <fact> *and* <formula> <formula value> *can be computed as* <formula>*."*.

According to ISO 5966 "Documentation – Presentation of scientific and technical reports" and ISO 7144 "Documentation – Presentation of theses and similar documents" formulas which are written in continuous text shall be written on one line (use $1/\sqrt{2}$ or $2^{-\frac{1}{2}}$ instead of solidus).

Important formulas, which play a **central role** for your Technical Report can also have a **legend**, where the influencing factors are explained. Here an example:

$$m \cdot g = \sigma \cdot d_n \cdot \pi \cdot f(d_n / \sqrt[3]{V}) \tag{3-5}$$

m – mass of the drop and its satellites
g – accelearation of gravity
V – volume of the detached drop
d_n – orifice diameter
σ – surface tension of the liquid
$f(d_n / \sqrt[3]{V})$ – empirical correction function, equals 1 for an ideal drop with no satellites and no material left on the tip

[23,26]

Formula and legend form an optical unit, which the reader can find again easily when he reads the report the second time or later. All required explanations to the formula (including the citation) are listed in the legend. The reader can – but does not need to – read the connecting text.

Reference to formulas: From the text you can refer to a formula by writing its number. The brackets are omitted: "..., see equation 4-5." Another example: "In equation 3-16 it shown, that ... ."

Computations of loads differ from the computations described so far, because there is **nearly no text**. Computations of load have outcomes like "$C_{required} = 23{,}8$ kN" etc.

Then there must follow a statement like: "selected: grooved ball bearing 6305 with $C_{existing} = 25$ kN $> C_{required} = 23{,}8$ kN". Here is another example:

2.4.3 Computation of loads for the arms of the puller

Load assumption: All four pullers bear equal loads

$$\tau_s = \frac{F}{A} = \frac{100\,000\ \text{N}}{50\ \text{mm} \cdot 40\ \text{mm}} = 50{,}00\ \text{N/mm}^2$$

$$\sigma_b = \frac{M}{W} = \frac{100\,000\ \text{N} \cdot 35\ \text{mm}}{\dfrac{50 \cdot 40^2}{6}\ \text{mm}^3} = 262{,}50\ \text{N/mm}^2$$

$$\sigma_V = \sqrt{\sigma_b^2 + 3\tau_s^2} = \sqrt{262{,}50^2 + 3 \cdot 50^2} = 276{,}42 \text{ N/mm}^2$$

selected: material S355JO

$$\Rightarrow \sigma_{allowed} = 355 \text{ N/mm}^2 > \sigma_{existing} = 276{,}42 \text{ N/mm}^2 \quad \Rightarrow \quad \text{OK}$$

In such computations the text stands back very much. There are only a few connecting words. The information is expressed nearly only in formulas and equations. This writing style may only be used in computations, which aim at finding numerical data.

All required formulas must be explicitly written, then the measures should be inserted – where useful with measuring unit –, and in the end there follows the result of the computation. This procedure is compact and reproducible. For a better overview in case of formulas which spread across more than one line, you should put the first equation sign of each line below the previous one.

For **general discussions, developments etc.** the writing style in the fracture mechanics example above (a few formulas, a lot of text) must be used.

3.6.4 Understandable Writing in Technical Reports

Understandability is a complex term in communication sciences. Whether a text is understood by the target group depends in general on two groups of properties. These are the "text properties" and the "reader properties".

Reader properties cannot be influenced by the author, since they depend only on the planned or random readers of the text. They are:

- general prior knowledge of the readers,
- their ability to concentrate on reading and their routine in reading,
- their motivation for or inner aversion against the topic(s) of the text,
- their command of the language in which the text is written,
- their knowledge of technical terms, and
- their expertise in the described field of knowledge.

The **text properties** are specified by the author beside a few exceptions like layout rules according to a corporate identity. Typical checklist questions referring to text properties are:

- Is the title descriptive and does it create interest?
- Is there a structure and is it logical ("backbone")?
- Paragraph length?
- Sentence length?
- Nested sentences?
- Are unknown words explained when they are used for the first time?
- Are unknown abbreviations explained when they are used for the first time?

- Are there too many foreign words?
- Has the reader all information available in all sections of the Technical Report, which he/she needs for following the contents?

This bullet list shows clearly, that the understandability of the text can be negatively influenced by an inappropriate design of the text properties above. Therefore we want to describe some possibilities to improve the understandability of non-fictional texts.

Improvements of the understandability of texts can be made on three levels. These are the *text level*, *sentence level* and *word level*, which are individually looked at in the following.

On the **text level** the elements that improve the understandability are at first the indices and lists, i. e. table of contents, index, glossary and lists of figures, tables, abbreviations, units and symbols etc.

Also ***headers and footers, marginalia*** as well as ***page numbers*** and ***cross-references*** improve the understandability on text level.

Moreover, ***introductory sentences*** at the beginning of a document part and ***connecting sentences*** at the end of a document part have to be mentioned here. These sentences can also be used for the introduction of figures and tables. They help the reader, to follow the "backbone", because they rebuild a connection to the logical structure again. They describe, which information follows next and why, how the following information must be evaluated, which significance the information described so far has etc.

Another measure, which improves the understandability of the text are ***footnotes***. Footnotes are superordinated numbers which can be created in Word with Insert – Reference – Footnote. Footnotes are consecutively numbered in the text. By default, the number appears behind the text it refers to and it is repeated at the bottom of the page below a left-justified, horizontal line of about 4 to 5 cm length. There you can place information, which would disturb the normal flow of reading. Examples for such information are

- comments about cited texts,
- bibliographical data of cited literature,
- the exact location of the cited information in the source of literature and
- hints or comments regarding the normal, continuous text.

In most cases (as in ISO 5966 and ISO 7144 and several other ISO standards) the footnotes appear in a smaller font size than the standard text. However, Microsoft Word formats them in the same font size as the standard text. (This is also recommended by DIN 5008.)

Footnotes are used quite frequently in the humanities[1]. In the technical sciences footnotes are less common.

---

[1]   Here is an example of a footnote, like it is usual in the humanities.

If one or more footnotes refer to the contents of a table, they appear directly below the table without a horizontal line of 4 to 5 cm. The German Association of Electrical Engineers recommends, that normal footnotes get superscript numbers and tables get superscript small letters to distinguish them from one another. You can manually create the superscript letters (Format – Character, Superscript).

For shorter texts, sometimes endnotes are used, beginning on the last page of the text. Endnotes can be converted to footnotes by pressing a button and vice versa.

The area of *typography (page margins, font type, font size, etc.)* brings also possibilities to improve the understandability of non-fictional texts, but only in a wider sense. In a wider sense, because the typography belongs more to recognizability or perceptibility than to understandability. A quite important rule in this context is: The wider the printing area is (or the longer the lines are), the larger the line spacing should be.

(Good) *figures and tables* open up the text, complete it and address other apperception channels than continuous text. So the author can display the information in several ways and on more than one level and the reader can follow the flow of information better. Therefore you should use many figures, tables and bullet lists as well as pictorial or tabular re-arrangements of text, see sections 3.3.5 and 3.4.10. Examples also make the text illustrative. However, too many examples can have a negative influence on the briefness and conciseness of your texts.

On **sentence level** there are also some rules which improve the understandability of non-fictional texts very much, **Checklist 3-13**. In addition, the translatability into foreign languages is improved, if you keep these rules.

Checklist 3-13 Rules for better understandability of texts on sentence level

- The sentences should be as short and simple as possible.
- Each new fact should preferably be described in a new sentence.
- Leaving out verbs to shorten sentences is not allowed.
- The sentences should not be longer than 25 to 30 words.
- Paragraphs should have maximum six sentences. Paragraphs with only one sentence should not appear too often.
- Tables and bullet lists should be used as often as appropriate.
- Compound tenses should be avoided (depending on the target group), the simple tenses (present tense, past tense and future tense I) are better understandable for most people.
- Abstract nouns (foreign words) are tiring for the readers and should therefore be avoided. This corresponds to the requirement to write Technical Reports concrete, not abstract.
- Meaningless phrases and filler words also appear as tiring, if they are used too often, and should be used carefully.
- If a word is used in an unusual meaning, please use quotation marks or italic printing.
- The first verb should not appear too far at the end of the sentence.

- Nested sentences should be avoided. Parentheses in such sentences should be only short. Their contents can evtl. be expressed in individual sentences.
- Double negations are in most cases superfluous. A normal negation shall not appear too far at the end of a sentence.

The sentence structure is very important for the understandability of the text. In general one main clause plus one or (more seldom) two sub clauses or a combination of two main clauses should not be exceeded.

The understandability of the text on sentence level can be improved very much by using conjunctions and disjunctions and some other means. With these ***conjunctions and disjunctions*** you design a logical structure of the individual sentence parts and connect the current sentence congruously to the previous one. Here are some examples:

not so easy to understand:
Cranes always have a maximum tolerable lifting load, which, if it is considerably exceeded, can result in a buckled crane boom, an overturned crane or broken lifting ropes.

better understandable by conjunctions:
*Cranes always have a maximum tolerable lifting load. **If** it is considerably exceeded, **then** the crane boom can buckle, the crane can overturn, or the lifting ropes can break.*

not so easy to understand:
The current I flows through the resistance R. It is relatively small.
Who is meant here? I or R?

better understandable due to clear references:
The current I flows through the resistance R. R is relatively small.

not so easy to understand:
The new sensor was much more linear than the old one.

better understandable due to precise expressions and facts:
The sensor characteristic of the new sensor has a deviation from a linear relationship between current and flow rate of max. 3 %. In case of the old sensor the deviation was max. 7.6 %.

☞ *Use clear, precise and significant formulations.*
Commit yourself. Write "white" or "black", but not "gray".

Expressions like rather complex, nearly linear, very fast, little power, highly sensitive and relatively low are much too vague and must be supported with figures, if they are important. If they are unimportant, they should be left out.

☞ *Try out to translate your text to a foreign language in your mind. If you do so, you will clearly recognize your complex sentence structures, unclear references and imprecise descriptions. You will probably wonder, how many alternative and more simple expressions and sentence structures you find, while you translate.*

In this way the author can influence the design and refinement of the information network, which continuously grows in the readers' brains while they are reading. The building of wrong information or wrong information references is avoided and the understandability of the text is improved.

On **word level** the amount of *simple, common, accurate words* shall be as high as possible. *Technical terms and abbreviations* shall be *explained* when they are used for the first time. In addition, technical terms and abbreviations can be explained in a glossary, abbreviations evtl. also in a separate list of abbreviations.

In general the number of used *foreign words* shall be *limited*. Technical terms sometimes consist of several nouns. In this case, you should use *hyphens between words which belong together*.

The following rule plays a central role for the understandability of text on word level:

☞ *Within one Technical Report the same appliance, assembly, part, fact or process is always consistently called by the same name, otherwise the reader is unnecessarily irritated!*

This rule also applies, if within one paragraph one part is addressed all the time. In the language courses they emphasize fluent speech. Sentence introductions, verbs and nouns are substituted by synonyms, so that the text does not become boring. In Word you can mark a word and look up synonyms with Special – Language – Thesaurus.

However, a Technical Report – other than a lyrical text – has to transport information without any ambiguity. Technical parts, facts and procedures may therefore be addressed *only* with their once specified names. What does the most beautiful formulation help, if the readers understand the message very much different from what the author intended to say? It is important that you do not presume too much prior knowledge. Authors overestimate the knowledge of their readers over and over again. Apply all suitable measures to make your texts clear, understandable and easy to read (figures, tables, bullet lists, intermediate headings etc.).

3.7 Using word processing and desktop publishing (DTP) systems

Today the usage of word processing systems is state-of-the-art. Since texts may be edited easily, some authors tend to write "something" to begin with. If at the end, when the project deadline is in sight, the time-pressure rises, necessary changes remain undone. If you look at Technical Reports you may see, from when on the time-pressure became very hard. Problems with inner logic, spelling, grammar as well as creating or integrating figures occur much more often from then on and the report creates the impression of being made "quick and dirty". To avoid this, you should have enough time for proof-reading, entering the corrections and for the "end-check" in your project plan.

If you create a Technical Report **as a group**, the group members can write separate parts of the report at home and combine these parts to one common file (or set of files) later. However, please care for a common typography in the individual files. For example, define a common page layout and common formatting templates before the group members start writing. Work on a common style guide for your Technical Report and exchange modifications of the style guide regularly.

Typography is the placement and distribution of printing ink on the paper. You can distinguish macro-typography and micro-typography.

Macro-typography is the design of a printed page, i. e. the page layout. This includes the definition of the page margins, the design of headers and footers, the placement of the page number, the design of table headings and figure subheadings etc.

Micro-typography deals with the design of the individual characters, i. e. the selection of the font type, font size, font attributes (superscript, subscript), text markings like bold, italic, underlined, etc. as well as the usage of space characters.

Now we want to give you some hints for the appropriate design of your report with a word processor, here Microsoft Word and Open Office Writer. For working with L^ATEX – a word processing system which is quite frequently used in universities – you have to pass an intensive learning phase. How to work with L^ATEX is described in the literature and it would exceed the framework of this book to go into the details here. A useful book regarding the work with L^ATEX is presented in the list of references of this book.

3.7.1 Document or page layout resp. and hints on editing

☞ *If you already use Microsoft Office 2007 and you do not find the functions any more, look here:*
http://office.microsoft.com/en-us/word/HA100744321033.aspx (Word)
http://office.microsoft.com/en-us/help/HA101490761033.aspx (PowerPoint)

When you define the **page layout** and the formatting templates, you have to keep the layout rules of your university or company regarding their **corporate design**. If there is a lack of rules in the corporate design guidelines, you may use the rules we propose here accordingly.

The definition of the **page margins** and the placement of the page numbers are flexible to a certain extent. The following rules have proven to be practical: Define the page margins right from the beginning of writing your Technical Report, so that the printing area is fixed and the line and page break do not change very much shortly before the deadline of your project. This includes, that in a group all "writing" members use the same settings and formatting templates, e. g. for page margins, header, document part headings etc.

Ideally you should speak with your copy shop or printer already at the beginning of your project, whether you provide the **printing data in digital form or whether paper originals** shall be copied. For digital printing you often need a special **printer driver**. You should install it already at the beginning, so that the line and page break do not change much any more shortly before the deadline of your project.

If the front and back sides of the pages of the copied Technical Report shall be used, you have to switch on alternating page margins in your word processor. In Word: File – Page layout – Margins, option "Opposite pages".

The **upper and lower margin** should be at least 20 mm wide on DIN A4 paper, if there is no header or page number. There should be a gap between header or footer or page number and the edge of the paper of at least 15 mm. The right/outer margin may not be narrower than 15 mm, 20 mm is better. The left/inner margin should be at least 25 mm wide in any case so that the text is visible also at the beginning of the lines.

When you define the page margins, you have to consider **how the Technical Report will be bound later**. If you use staples, plastic or cardboard folders, plastic spiral (comb) binding, wire-o-binding (with wire spirals) and staple binding you need a wider inner margin compared with an adhesive binding. Also, if the size of the pages of the report is to be **reduced with the copier** (e. g. to DIN A5), the margins on the DIN A4 original must be wider.

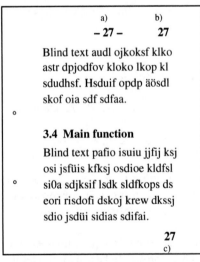

Usual position for the page numbers
(a and b are recommended)

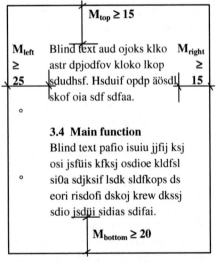

Size of the page margins in mm

When **entering the page numbers** you have to distinguish at first between book page numbering and report page numbering. In books the front and back side of the page are used, while in Technical Reports usually only the front side of the pages is used. The page numbers can be placed in the middle of the top margin, on the right side of the top margin, or on the right side of the bottom margin.

If the right and left pages shall be used, the page number must always appear on the outer edge of the pages and you need just and injust headers (in Word use option "Just/injust differ" in menu File – Page setup – Page layout or tab Page margins). In Open Office Writer you use different formatting pages to achieve this. On the right side there are always the odd page numbers (1, 3, 5, 7 etc.).

The design of **headers and footers** shall be decent. Headers which expand across more than one line – evtl. even with a logo – often appear to be overloaded. In the **header** you can show the chapter heading and the page number. A thin line (underlining), an appropriate distance to the normal text or a different font style (e. g. italic) as well as a smaller font size can be used to distinguish the header and footer from the ordinary text. If you use a thin line, it should start at the left edge of the printing area and end at the right edge of the printing area, i. e. the thin line is exactly as long as the printing area is wide.

If the front and back sides of the pages shall be used and in the header there is the chapter heading on the left and the subchapter heading on the right page, this is called **column titles**. The headings in this book are an example for these column titles.

In the **footer** – also divided from the normal text by a thin line – you can give information about the version of the document and/or a copyright note.

The example pages above show, how and where page numbers can be placed and how the margins can be defined.

Up to now we have introduced recommendations for the margins and headers and footers. Now the text in the printing area shall be structured and designed by typographic means. In Word such settings refer to a paragraph, if they are not assigned to single words or characters via Format – Font. Therefore they call it paragraph layout. The **paragraph layout** can be controlled by the formatting functions or by applying the paragraph formatting templates. In general, the standard font size should not be smaller than 10 pt.

There should be a vertical spacing of ≥ 6 pt up to one line. You should use the function "Format – Paragraph – Distance, Below:".

For **document part headings** you should take as a rule of thumb, that above the heading there should be two blank lines and one blank line below it. In any case the distance above the document part heading shall be distinctively larger than the distance below it, so that the reader can recognize more easily, that a new document part starts here and which document part the heading belongs to, see example pages above. For figure subheadings and table headings this rule should be applied accordingly, see 3.3.2 "Table numbering and table headings" and 3.4.2 "Figure numbering and figure subheadings".

Document part headings are printed in boldface in most cases. A larger font size compared with the normal text is recommended. Depending on the hierarchy level of the document part heading there are different font sizes for the headings.

The gap between document part number and document part title should be at least two space characters wide. If you create the tables (of contents, figures, tables etc.) automatically, this gap should be created by a tab. Then the tab is copied from the document part heading in the text to the automatically created table of contents and you can layout the table of contents more easily.

Figure subheadings and table headings are layouted in such a way, that the figure or table title can be read and their labels can be distinguished from the normal text quite easily, because in cross-references to figures or tables in the normal text these labels are used as search criteria. If you look for the figure/table to which the cross-reference refers, the information: "here is a figure/table" comes out clearly. To identify the figure or table *number* more easily, bold print of the label (like "Figure 12" or "Table 17") within the figure subheading or table heading has proven to be practical. The label is distinguished from the title of the figure subheading or table heading by two space characters or – better – by a tab. A colon behind the number is no longer usual. Here are an example for a figure subheading and an example for a table heading.

Figure 12   Principle of under-powder welding

Table 17   Filter weight depending on the welding position

Figure subheadings or table headings which spread across more than one line are layouted so that all lines of the figure or table title start at a common **building line** (use tab or hanging indentation!).

The **line spacing** "1 ½" is a good value for Technical Reports (in Word use the menu Format – Paragraph – Line spacing). Today they also use other (smaller) values for the line spacing. However, the eye can hold the line better, if the spacing is 1 ½. Very bulky works can become too thick and unhandy with this large line spacing. To avoid this, you may use both sides of the pages for printing or apply a smaller line spacing.

Now there are some remarks regarding the **text justification**. Normal text is either printed left-justified or – more often – left- and right-justified in the Technical Report. The left- and right-justification emphasizes vertical lines. This strengthens the optical effect of indentations and building lines. Moreover it looks more pretty. Figure subheadings and table headings may be centered, if the figures and tables are also horizontally centered. The title of a short article may also be centered, if there is no title leaf.

Indentations may be created in different ways. For example the following operations result in an indentation:

- inserting space characters (however, this does not result in a defined indentation in case of left- and right-justification!),
- defining left-justified tabs,
- moving the left indentation markers in the current paragraph (hanging indentation),

- defining an indentation via menu (in Word Format – Paragraph – Indentation),
- using a table without borders and leaving the first column empty.

In practice there are often texts which use many different options to create an indentation within one document and which have very different indentation values. The result is a layout which appears untidy and when you enter text corrections you have to find out first, with which method the indentations have been created. Therefore you should use only a few methods of creating indentations and only a few indentation values (e. g. tabs at 5, 10, 15, 20 and 25 mm).

List structures can be automatically created (e. g. in Word) with the two icons Numbered list and Bullet list in the toolbar Format. You can control their appearance in the menu Format – Bullets and Numbering. There you can use the preselected bullets or press the button Customize… and select different bullets. Creating nested lists with different bullets is now possible with a simple mouse click, if you use the indention buttons next to the list buttons . With Word 2002 this was not so comfortable and therefore, the authors have created own paragraph format templates, e. g. Level2 for the second level.

The selection of the **font type** is also a very important decision. Fonts from the **Times** family have proven to be adequate for larger amounts of text. Due to the serifs (the small lines at the ends of the letters) the reader can hold the line well during reading. When he switches to the next line, his eyes do not erroneously jump to the over next line. The reader is well-accustomed to these fonts from reading newspapers or books. Fonts without serifs (e. g. **Arial, Helvetica**) can be used for title leaves and overhead slides.

The normal fonts are **proportional fonts**. The distance from the end of one letter to the beginning of the next letter is constant. These fonts are not so well suited for the design of figure tables, where it is necessary, that the figures have the single, tenner, hundred and thousand digits aligned one below the other. For example, the digit "1" is narrower than the other digits. A **font with fixed spacing** (e. g. Courier, Lucida Console, Monospaced) might help in this case. Here the distance from the middle of one letter to the middle of the next letter is constant. If you design tables with these fonts, you will not have to use tabulators so much. Simply adding space characters is enough to create common building lines for tables, bullet lists, equations etc. The tables and bullet lists are then easier to read. This is especially the case for long tables and computer listings.

For the **font size** the following values should be applied. The usual fonts from the Times family should not be smaller than 9 pt. This is the lower limit of readability. A font size of 10, 11 or 12 pt is well readable. The standard font size should not be larger. If the size of the Technical Report shall be **reduced to DIN A5** with a photocopier, the font size should be 13 or 14 pt for the standard text.

Text accentuations can be achieved in different ways: **bold**, *italic*, underlined, ***bold and italic***, SMALL CAPS, CAPITAL LETTERS (majuscules), but also by frames or a fill.

These accentuations have different functions. They can show literature citations [MILLER, 1989, p. 151], emphasise **important parts of the text** or highlight an *un*usual sense. Also, "technical terms" may be denoted, or via (an annotation or) the insertion – idea – of thoughts words or text passages may be accentuated. However, the text accentuation also disturbs the flow of reading. Depending on the text type you should apply a different amount of text accentuations:

- The **classic book layout** allows only little text accentuations. If they cannot be avoided, in seldom cases italic print is allowed.
- **Advertising texts** are the other extreme. They are created to get and hold the attention of the reader and to position the message "buy me" as deep as possible in the readers heart and mind. They address the human psyche in a clever way and apply unusual effects of text and image design, create new words, and exceptional sentence structures to achieve more turnover. Everything is allowed, which is successful and not displeasing.
- **Instructional texts** (user, maintenance, repair, installation or operation manuals, training and seminar documentations etc.) use text accentuations for attracting and controlling the readers' attention (like this book, which is therefore a little different from the classic book layout). Text accentuations use the reading conventions of the people to display information so prominently that they cannot be overlooked. However, too many text accentuations are too much of a good thing. If too many information pieces are marked as important by means of text accentuations, the reader cannot decide any more, which information is really the most important one. He/she is irritated and frustrated.

You as the author of a Technical Report have to find an acceptable balance. You have to keep in mind your **supervisor or customer and** the **target group**. The central question is: How much creativity and unconventional design is acceptable?

- Are the supervisor or customer and the target group conservative? Then select a simple and decent layout! Technical information is assumed to be much more important than a functional and clear design.
- Are the supervisor or customer and the target group informal and unconventional? Then the text may be formulated informally and more text accentuations are allowed!
- Are the supervisor or customer and the target group top managers? Then you should do anything you can to shorten the text and present important information in graphics.

Moreover, it is **important, how the text is probably read**. Is it read like a textbook **sequentially or** like an instruction manual or an encyclopedia **selectively**? If you assume, that the text is read in a selective way, it is extremely important to work with many text accentuations. Beside the already mentioned text accentuations with different printing (bold, italic etc.) there are the following options:

- **Marginal notes**
  At the outer margin of the document in an own column there are keywords which represent the contents of the appertaining paragraph. Well-suited for introductory non-fictional texts.
- **Register**
  At the outer edge of the pages there are black or colored marks. This helps to distinguish different chapters. Well-suited for encyclopedias and product catalogues.
- **Column titles**
  Well-suited for alphabetically-ordered information. At the upper margin of the document on the left side there is the first and evtl. on the right side the last alphabetical entry on the current page. Example: Telephone book or encyclopedia.

The following questions are essential for the application of all text accentuations: Which information must be found how fast and secure? Which search strategy do the readers apply? Which reading conventions can be assumed in the target group? Moreover, even a very sophisticated system of attracting and controlling the readers' attention may not forget the following basic condition:

☞ *To achieve that the readers can use the usual lists and tables and marginal notes at all (table of contents, index, list of figures, list of tables etc.), they must know the words being used there and search for them themselves! This means, that these entries must be **answers to questions asked by the readers** (not by the author).*

In section 3.7.3 "Details about text accentuations" **Table 3-12** summarizes the typographic possibilities of text accentuations and concrete rules for their application.

Now there are some rules for a good **page make-up** There are some information units, which must never be separated by a page break.

- A document part heading may never stand alone or together with only one text line at the bottom of a page.
- A single text line of a paragraph may neither stand alone at the bottom nor at the top of a page, unless the paragraph consists of only one line.
- A table heading and the appertaining table may not be separated from each other by a page break. Also, the individual rows of a table may not just be splitted. In this case the table must be continued on the next page with the same table heading and a continuation note. Alternatively, you can move a paragraph above the table which stood below the table or vice versa.
- This is also true for figures and figure subheadings.

If possible, **figures** shall **be located on the same page as the appertaining text** or on the following page, so that the information in the figure is presented *near the text*. This improves the text-figure-relationship and the understandability of the Technical Report. If you want to avoid that the figure is positioned on the next page, you can only shorten the text, move a paragraph below the figure, use smaller empty lines (e. g. 10 pt instead of 12 pt) or apply the function Format – Paragraph – Distance below (a paragraph).

At the end of a chapter the **last page before the next chapter heading** is quite frequently not completely filled. If in such a case on the last page of the preceding document part the page is at least filled by 1/3, this is acceptable. Otherwise some paragraphs can be moved from one file to the next or the text may be shortened or the size of figures may be enlarged or reduced, so that the page break fits better.

Basically it is clearer, if each chapter begins on a new page. If both sides of a document are printed, sometimes a new chapter may only start on the right side where the pages have odd page numbers. In this case the last page of the preceding chapter – on the left side – may remain empty. In the Technical Report in most cases the report page numbering is applied and the back sides of the pages are not printed. In this case it is best, if each chapter begins on a new page (1st hierarchy level in the structure!).

The **line break** may also be influenced systematically. The term line break contains all measures of an author which result in a new line. In document part headings, figure subheadings and table headings it may be undesired that a word is automatically hyphenated. Then you can create a line break with the key combination Shift + Enter. Also in bullet lists it may make sense, to enforce a new line in this way, so that the reader can read the information in specific information blocks controlled by the author.

If the text is left-justified and right-justified, you have to enter a **tab before the enforced line break**, so that the last line before the line break is left-justified and large gaps between words are avoided. If you switch on the display of non-printable characters (¶) this looks like: → ↵.

In the continuous text the line break can mainly be influenced by **inserting hyphenation proposals (optional hyphen, soft hyphen)**, if the automatic hyphenation results in undesired or wrong hyphenation or if due to missing automatic hyphenation there are very large gaps between the words (in extreme cases even between the individual characters). To influence the hyphenation, there are special characters. These are non-breaking space, protected hyphen and soft hyphen. In Word you reach these characters in the menu Insert – Symbol (or Special character) – tab "Selection 2".

The **non-breaking space** is used between the components of a multi-part abbreviation or between abbreviated titles and family names. It creates a constant word distance, which may be smaller than between normal word distances, if the text is left-justified and right-justified. Also the non-breaking space prevents the automatic hypenation, so that names like Dr. Mayworth, abbreviations like e. g., i. e., figure or table labels like Figure 26 as well as value and physical unit like 30 °C, 475 MPa etc. are not divided, but stand together on the next line. In Word you create a non-breaking space with the key combination Ctrl-Shift-Spacebar.

The **protected hyphen** shall – similar to the non-breaking space (or protected space) – circumvent a hyphenation. It is e. g. used in words combined of a letter and a word like e-mail, X-rays etc.

If you enter hyphens, you should insert **soft hyphens** (key combination Ctrl and Minus in the normal keyboard). If you enter normal (hard) hyphens (only Minus in the normal keyboard) and insert or delete text later on, the hyphens appear somewhere in the middle of the line, which should be corrected, but which is often forgotten or overlooked. In this context look carefully at newspapers and journals. There this error also occurs quite frequently.

Sometimes there are terms consisting of **combined words**, which can be connected with a hyphen. This improves the readability and understandability of your text. Please use such hyphenation consistently throughout the whole Technical Report. Example: "rubber block drive belt" becomes "rubber-block drive-belt".

While editing text, you are sometimes astonished about **automatic functions**. They shall help to avoid frequent mistakes, but sometimes they are very annoying. If such an automatic function creates undesired results, you should look in Word how the options are set in the menus Format – AutoFormat and Tools – Auto-Correct Options. By the way, in the AutoCorrection table you can delete undesired entries and enter your own, typical typing errors, if they shall be automatically corrected. Also quite interesting are the tabs "Compatibility" and "Spelling and Grammar" in the menu Tools – Options. You should read carefully there. In most cases, there is an option which controls the undesired effects, which can be switched to try out working with different settings than the standard settings.

It is generally recommended, to format larger documents with **formatting styles** instead of individual assignments for each paragraph. This is true for all **frequently-used layout patterns** like document part and table headings, figure subheadings, bullet lists, cited text etc. Indentations should not be created with space characters, but with tabs or indentations in the Format – Paragraph menu. Such an approach saves work and assures a uniform layout.

Automatic cross-references also save work, see 3.7.4 and 3.7.5.

The information about the document or page layout and the paragraph layout have mainly dealt with the problem to distribute the "printing ink globally on the paper" (macro-typography). Now we want to present some general information and rules about entering the individual characters (micro-typography).

3.7.2 Typographic details according to good general practice

While in manual writing the distances between words, punctuation marks, formula signs and digits have only limited importance, this changes in writing with the computer. There a space character creates a well-defined gap. In the following we present a short summary how the space character, brackets, calculation symbols, values and physical units, numbers etc. must be used, **Table 3-10**.

Table 3-10   Usage of space characters, brackets, calculation symbols, values, physical
units etc. <to be continued>

Example	Rule
… is finished. Therefore … … stressed, so that …	**The following punctuation marks** are used <u>without</u> space character directly after the preceding word: . , ; : ! ? After them there is always one space character.
cross-reference	Also the **hyphen** is connected to the preceding and following word <u>without</u> space character.
Due to cost reasons, ball bearings must be examined first – although cylinder roller bearings have a much higher load bearing capacity.	Before a **dash** there is always one space character. After a dash there is either a punctuation mark (like comma or period) or a space character and more text. You can insert a dash via Insert – Symbol (or Special character), font type: Symbol.
… (text) … … [text] … … {text} … … <text> … … /text/ …	**Brackets** directly enclose the text without space character between bracket and text. This is also true, if you use slashes to mark literature citations.
80 – 85 m, 65 – 70 kg 4,– to 4,80 EUR 09:30 – 13:30 h but: from 09:30 to 13:30 h Elm Street 10 – 12	The sign between two values which limit a value range is a dash or the word "to". The word "to" is recommended to avoid mismatches with the calculation symbol minus.
… [17], this is insofar … according to MILLER [16]. WINTER [19] has watched, that …, while SHARK [20] presents completely different test results.	Before a citation [17] (or /17/) there is always one space character, regardless whether it occurs in the text or in a figure subheading or table heading. A period at the end of a sentence or comma to connect a main clause and subordinate clause is added without, normal text with a space character. An ellipsis is separated from the text by a space character before and after the ellipsis. A comma follows without space character. The ellipsis evtl. replaces a period at the end of a sentence.
3 3/5 Winter semester 1995/96 and/or	If the **slash** is used in a fraction or in the sense of "and" or "or respectively", there is no space character before or after the slash.

Table 3-10 Usage of space characters, brackets, calculation symbols, values, physical units etc. <continued>

Example	Rule
47 cm 16 kg 26 °C	Between measure and physical unit there is always exactly one space character (or non-breaking space). Exception: angle degrees.
16° or 16 degrees 5°2' or 5 degrees 2 minutes	In angle degree specifications there is either a degree symbol directly behind the measure or the word "degrees" is added with a space character (the same applies for the words minutes ' and seconds ").
CO_2 10^{-3} n_0 $(x+12)^2$	Exponents and indices are added to the base without space character.
$2 + 5 = 7$ $6 : 3 = 2$ $-5 \times 3 = -15$; $-5 \cdot 3 = -15$	Before and after calculation symbols there is exactly one space character. Algebraic signs appear directly before the figure.
i. e., e. g., evtl., U. S.	Abbreviations, which are spoken as the full word(s) get a point. This point evtl. replaces a period. Several abbreviations following each other are separated by a space character (or non-breaking space).
NATO, UNESCO, USA, EU	Abbreviations which are read as letters or as an own word are written without points and without space characters between the abbreviated letters.
The delivery of the ordered ball bearings follows until August 12, 2009. If you are dissatisfied with the quality of the bearings, please contact the address above.	**Between two paragraphs** there is exactly one empty line.
50.22 127.85 1,295.33 23.10	**Groups of figures** are aligned according to the decimal point. If there is none, they are right-justified. Long figures (more than three digits before the decimal point) can be grouped with a space or non-breaking space or comma in groups of three digits each. A figure with non-breaking spaces cannot be divided by a line break. Due to safety reasons **amounts of money** shall be grouped with commas and the decimal point.

Table 3-10   Usage of space characters, brackets, calculation symbols, values, physical
               units etc. <continued>

Example	Rule

The goods your ordered on March 30 contain:
 	
– 500 Washers
   ISO 10510-St 6.3
– 500 Hexagon head bolts
   ISO 4014 M6 × 50-8.8-A2P
– 500 Hexagon nuts
   ISO 4032-M6
 	
The delivery follows until May 15, 2009.
	Enumerations are layouted as bullet lists and separated from the continuous text with exactly one empty line before and after the bullet list. The enumeration items (especially if they have more than one line) can also be separated from each other by exactly one empty line.

These rules apply for any computer-written text including letters, e-mails and Technical Reports. These rules are e. g. standardized in DIN 5008. Besides there are some more typographic rules which are not standardized, but which have become popular and should be adhered to, **Table 3-11**.

Table 3-11   Common abbreviations and notations

Example	Explanation
€, EUR	Euro
kN	Kilo Newton (1 000 Newton), the small letter "k" is used for all units beside Byte
KB	Kilo Byte (1 024 Byte)
t/a	tons per year
&	The ampersand (et-symbol) should only be used in company names, it is not allowed in continuous text as a substitute or abbreviation for "and". This is also valid in presentation slides. Therefore the following is wrong: "... against falling out of persons & devices ..." It should be: "... against falling out of persons and devices ..."

3.7.3 Details about text accentuations

Text accentuations are typographic means to emphasize certain figures, words or phrases compared with the continuous text. They have already been addressed a little in section 3.7.1 "Document or page layout resp. and hints on editing". Now we want to summarize the hints and give examples for their usage, **Table 3-12**.

Table 3-12   Typographic means for text accentuations with examples of their usage
                      <to be continued>

Typographic means	Fields of use
bold	document part numbers and headings, titles, figure and table labels, **strong** emphasis compared with continuous text.
italic	in formulas: for scalar values like U, I, F, l, D, z, *weak* emphasis compared with continuous text, is also used to optically distinguish page headers and footers as well as figure subheadings and table headings from the continuous text.
CAPITAL LETTERS	Headings (but capital letters are harder to read than capital and small letters), identification of Author names in continuous text, emphasis in ASCII files, loudly shouted/yelled statements in e-mails.
SMALL CAPS	Headings written in capital letters are harder to read than capital and small letters), identification of Author names in continuous text.
underlined Tips for underlining Venue: Townhall Time: 05-17-2009, 16:00 h Frequences (Radio, TV, CB radio etc.)	medium emphasis, – stronger than *italic* and – weaker than **bold**, The underlining goes from beginning to end without gap! A colon in announcements is also underlined. Additions in brackets after headings are not underlined.
__important__	Substitute for underlining in text-only e-mails
double underlining	Identification of final results in formulas and final sums in computations.
indented	Identification of longer, literal citations, mnemonic rules etc. Indentations must not be too far, 5 or 10 mm are often sufficient.
shaded (with fill)	Identification of variants, formulas, headers and introductory columns of tables, distinction right/wrong, before/after etc.

Table 3-12   Typographic means for text accentuations with examples of their usage
                  <continued>

Typographic means	Fields of use
framed	Examples, mnemonic rules, formulas, summaries, link lists.
other font type	Font for continuous text = Times New Roman special information type = often other font type *Examples:* Computer dialogue: `Courier New` Dialog field display Monospaced in a user manual: Font on posters Arial and overhead slides: Formula symbols: Symbol (αβγδε)
other font size	Headings larger than standard text, depending on the hierarchy level, figure subheadings and table headings evtl. smaller than standard text

The different typographic means shall be used **consistently**, i. e. uniformly. Take notes in your style guide, **which type of information shall be displayed with which typographic means**.

3.7.4 Automatic creation of indexes, tables, lists, labels and cross-references with Word

Due to our knowledge several users, who use Word only seldom, do not know exactly how to automatically create indexes and tables with it, especially if the tables shall be created from more than one single file. Online help and user manual do not provide sufficiently clear assistance and the function central document – branch document is seldom recommended. However, in Technical Reports you often need to create tables automatically. Therefore the required procedures will be explained here. The following list is a step-by-step instruction what to do. At the end of this section there will be a list, how to create labels (like figure and table numbers) and automatic cross-references with Word.

Preparations

- Open a new file, e. g. "lists.doc". Switch on normal view and the display of non-printable characters with the function "show/hide hidden text" ¶ . Non-printable characters are e. g. the paragraph end marks, the symbols of space characters, index entries as well as comments, which are formatted as hidden text with the function "Format – Font".
- Please take into consideration that the line and page breaks change, if you switch the hidden text on or off.
- Now tell Word, which files shall be used for the creation of indexes and tables. Select "Insert – Field – Categories: Index and tables – Field names: RD". Please read the description in the function window. It says "Creates an index, a table of contents or a list of tables from several documents." If you click on the OK button, you insert an empty RD field. Select this field with the mouse and make sure, that the limiting brackets are both selected (in the following example these brackets are underlined). Copy this RD field with Ctrl-C and Ctrl-V so often, that there is one RD field for every text file which must be included. Then enter the file names into the RD fields, e. g.
  { RD 1-intro.doc } { RD 2-plantrep.doc } { RD 3-createtrep.doc } etc.

Table of contents

- **Fehler! Textmarke nicht definiert.**If you want to automatically create a table of contents, you have to format chapter, subchapter and section headings with the appropriate formatting templates. It is best, that you use the standard formatting templates for headings "Heading 1, Heading 2, Heading 3 etc." and define their layout (optical appearance in the text) according to your needs. During the automatic creation of a table of contents from formatting templates by default the paragraphs which are formatted with "Heading 1, Heading 2, Heading 3 etc." are copied to the table of contents and formatted with the formatting templates "List 1, List 2, List 3 etc.".
- If you do not want to use the standard formatting templates "Heading n", you should select "Insert – (Reference –) Index and tables, tab Table of Contents, Options" and select different formatting templates.
- You are in the file "lists.doc" or, if your report consists of one file only, at the beginning of the file, evtl. after the title leaf and front matter. In Word open the tab Table of Contents as just described. If you click on the OK button now, a table of contents is created from all available files and inserted at the current cursor position.
- The newer Word versions format the table of contents with the formatting templates "List 1, List 2, List 3 etc." and then in addition assign the headings with the formatting template "Hyperlink". This formatting template does not change the text properties like e. g. font type, font size and indentions. To influence the text properties, you still have to edit the formatting templates "List 1, List 2, List 3 etc.".
- If you click somewhere into the newly created table, the text appears in light gray. If you insert empty lines and apply other paragraph formatting the page break can be optimized and the layout can be further refined. Edit the automa-

tically created table like normal text. For example replace the tabulator with leading dots by "space character, tabulator, space character". Then optimize the line break. If you notice a typing error, you have to change this in the relevant document part heading in the chapter files.

- For the design of the document part headings please refer to section 3.1.2 "Structure with page numbers = Table of Contents (ToC)".

List of Figures (LoF), List of Tables (LoT)

- You are in the opened file "lists.doc", evtl. behind the just created table of contents or, if your report consists of one file only, in the appendix. Enter a page break and e. g. the heading "Figures" or "Tables".
- A list of figures or list of tables is created in a similar way as a table of contents. Format all figure subheadings with a newly defined formatting template like "FigSubheading" and all table headings with "TabHeading". With these formatting templates you define, how the figure subheadings and table headings shall look like in the text.
- You may change the appearance of the created list of figures and list of tables as follows: "Format – Styles and Formatting – click on the formatting template ‚List of Tables' – Edit". By default this formatting template is not contained in the normal formatting template list. Select "Display: All formatting templates", so that you can select and edit this formatting template.
- The newer Word versions in addition assign the entries in the list of figures with the formatting template "Hyperlink". This formatting template does not change the text properties like e. g. font type, font size and indentions. To influence the text properties, you still have to edit the formatting template "List of Tables".
- If desired, you can insert a page break in the file "lists.doc" before creating the list of figures.
- Select "Insert – (Reference –) Index and Tables – tab List of Figures". Under "Formats" select e. g. "Elegant" and under "Fill character" leading dots. Under "Options" you can tick "Formatting template" and select the relevant formatting template from the list, e. g. "FigSubheading". If you click on the OK button on the tab now, a list of figures is created.
- If Word asks, whether it shall delete the existing list, answer with No. As a result you will then have a table of contents and a list of figures in your file "lists.doc". Now you may copy the final list of figures to the correct position in your Technical Report, e. g. into the chapter appendix.
- If you want to create a list of tables, on the tab List of Figures under "Options" just select the formatting template "TabHeading". Then continue in the same way as for the list of tables.

Index

- Index entries must be inserted into your text files at the position they refer to. If you insert an index entry, Word automatically changes to the mode "show hidden text".
- First, define index entries in your text files. Mark text, that shall appear in the index and select "Insert – (Reference –) Index and tables, tab Index – Define

entry". The marked text is prepared as index entry. Now you can change the case of the letters, the orthography, singular or plural etc., so that the entries are well structured and fit together. Click on "Define", then back into the text window and mark another entry. If desired, you can also type an index entry into the edit field in the function window and click on "Define".

- Word inserts an XE-field into your text at the cursor position for each entry you have defined. Alternatively you can also use the function "Insert – Field". You can create subentries in up to 9 levels by entering a colon into an XE field (either from the function window or directly in the XE field). The subentries are indented in the index.

- The text between the quotation marks in the XE field may be up to 64 characters long. Main entries can be emphasized in the index by printing the appertaining page number in bold or italic printing. To achieve this you have to tick the desired character formatting in the function window or enter a command into the XE field behind the closing quotation mark and a space character: "\b" (bold) or "\i" (italic).

- If the index entry in the text is formatted in bold print, in the index not only the page number but also the index entry itself will appear in bold print, too. The same is valid for character formatting like italic, bold and italic, underlined etc.!

- If all index entries are ready, go to the end of your file "lists.doc" or of your report file. There you enter a manual page break again and the heading "Index". To create the index select "Insert – (Reference –) Index and tables, tab Index – Type: Indented – Formats: Formal – Columns: 2" and click on the OK button.

- You can influence the appearance of the index, if you edit the paragraph formatting templates "Index 1" to "Index 9". Details on the formatting template assignment are described under Table of Contents (ToC) and List of Figures (LoF), List of Tables (LoT).

- The newer Word versions in addition assign the index entries with the formatting template "Hyperlink". This formatting template does not change the text properties like e. g. font type, font size and indentions. To influence the text properties, you still have to edit the formatting templates "Index 1" to "Index 9".

- Word will only create the index in the final size, if the settings for page margins and headers and footers in the file "lists.doc" are identical with those in the other text files.

- A multi-column index is displayed in one column in normal view. Therefore you have to change to page view to look at the final index. Here you can optimize the page breaks.

Glossary, List of formula symbols and units etc.
- These lists do not serve to find information at a specific location in the text, but they have the characteristics of an encyclopedia. Therefore in these lists there are no page numbers listed. They are not created automatically, but manually typed like the list of references or normal text.

- If you would like to get a list with page numbers for your personal overview, you should apply the same work steps as for the list of figures and list of tables and use formatting templates like "Glossary", "Formula", "Rule", "Citation" etc.
- Eventually you want to insert hidden labels for your glossary entries, formulas, rules, citations etc. into your text files. You should write the label text into an empty line, mark the label text, and use "Format – Font, Effects: Hidden".

Correct page numbers
- Word only inserts the correct page numbers into your indices, tables and lists, if in each text file the next following page number is entered at "Insert – Page Numbers – Format – Begin with".

Usage of the indices, tables and lists to check your own work
- You can print or edit the automatically created lists like normal text – as described under Table of Contents (ToC). That means, you can copy these lists from your file "lists.doc" and insert them into your text files at the desired locations.
- To create these files is not only interesting for the final printout. You may already use these lists during the creation phase of your Technical Report. To have an overview which figures you have used at all, on which page they are and whether they have the correct page numbers, you can create a list of tables and list of figures from time to time. The same is true for the table of contents and all other indices and lists.

Too little main memory
- If your main memory (RAM) is not sufficient to keep all files open, which are needed to automatically create the indices, tables and lists, you have to execute the creation of the indices, tables and lists in consecutive steps. For example, create the following files:

  - "lists1.doc" with { RD chap1.doc } { RD chap2.doc } { RD chap3.doc } and
  - "lists2.doc" with { RD chap4.doc } { RD chap5.doc } { RD chap6.doc } etc.

The **complete table of contents** is then combined at one common location via copy&paste from the tables of contents created in lists1.doc, lists2.doc etc. The **same** procedure can be applied accordingly **for the other lists** (LoF, LoT) but the index. A **complete index**, which consists of the indices created in lists1.doc, lists2.doc etc. must be alphabetically ordered in the end. Change to normal view and order the index entries manually or use the **following trick**:

  - Delete all empty lines, copy the lines with the initial letters to the beginning of the file.
  - Select all index entry lines.
  - Select Table – Convert – Text to Table, keep the standard settings and click on OK.
  - Both table columns are selected.
  - Open Table – Sort, keep the standard settings and click on OK.

  – Select the table and use function Table – Convert – Table to Text.
  – Copy the lines with the initial letters to the right positions and insert empty lines.
  – Then select the whole text, open Format – Columns and select two columns.
  – The last step is to copy or cut and paste the complete index to the desired location.

Automatically created labels in captions (headings, subheadings)

Automatically created labels save time. Example: Number the figures in the figure subheadings with "Insert – (Reference –) Caption, Category = Illustration". New, own categories like e. g. "Figure" can be created, if you click on the button "New label" (or in earlier versions "New category..."). Then the drop list under Options also contains "Label = Figure" (or "Category = Figure"). Here is an example for a label, which is created with the category "Illustration":

$$(a+b)^2 = a^2 + 2ab + b^2$$

Illustration 1: First binomial formula

Other categories offered in Word by default are "Equation" and "Table". Automatically created equation numbers appear below the formulas and not, as usual, on the right side of the equation.

Automatically created cross-references

You can insert a cross-reference with "Insert – (Reference –) Cross-reference, Reference type = Illustration". Other available illustration types are e. g. text label, equation and table. In the field "Refer to:" you control, what is displayed in the cross-reference. Here are some examples, how the cross-reference fields can look like, the automatically created part is underlined.

Complete label:	Illustration 1: First binomial formula
	(it is impossible to print in bold only **Illustration 1:**)
Only category and number:	Illustration 1
Page:	171
Heading text:	3.7 The usage of word processing and DTP systems

If you have created labels you can refer to them in the text of your Technical Report with a cross reference. Example: see Illustration 1 on page 171. There are other options in the field "Refer to:" for other categories.

List of figures or list of tables with automatically created labels

If you want to create a list from the automatically numbered objects, open the menu "Index and tables, tab List of Figures". In the Options do not use the formatting template to be used but the category and click on OK.

Updating fields, indices, tables and lists

Indices, tables and lists which have not been automatically updated contain nearly always errors (non-conformities between text and indices, tables or lists). The most important advantage of automatic creation of labels, indices, tables and lists is, that the entries in the text and in the lists are identical and the identity must no longer be individually checked by individually turning the pages during the end check (as long as you did not change something in the text or the list after the last automatic creation of the list!). You update all automatically created fields and lists by pressing **function key F9**. This should be the first operation after opening your Word files during the end check.

3.7.5 Text editing with OpenOffice Writer

If you use OpenOffice.org Writer, which is available for Linux and Windows, you can edit texts in a similar way as if you use Word. The program has many nice presettings. If you are not used to the program, the online help is an important companion. It explains quite nice the usage philosophy of the different functions. The stability and reliability of the program is notable, even for documents which consist of 350 pages and more.

A useful help is the **formatting template toolbox**, which can be opened with **F11**. There are individual tabs for paragraph, character, frame, page and list formatting templates. To change the settings, click on an existing template with the right mouse button. A context menu appears with the alternatives New and Change. In the next window there are tabs for diverse settings again. You may delete templates created by mistake using the Del key.

Paragraph and character formatting templates can be used like in any other word processing or desktop publishing system. Two options on the Position tab are interesting: text can be rotated by ±90° and pair wise kerning (optimized character distance) can be used to optimize the readability of the text.

Frame templates are made for objects, which you would position with a text frame in Word. Here you find settings for borders (in general), graphics, OLE objects, formulas, marginals, watermarks and labels. There are already defined many templates with mostly quite intelligent settings. Frames can be positioned either centered in a paragraph or centered on a page.

Page templates help, if you want to use different headers in each chapter. You should define individual page templates for each header. For example, you could create a page template "Chapter 1", which is shown in the header of chapter 1 and another page template for "Chapter 2".

Here is the procedure to create a **new page template**:

1. Open a new text file, select Format – Styles and Formatting and click on the page template icon.
2. Click on the icon New template from selection.
3. In the field Template name enter a name and then click on OK.

4. Double-click the name in the list to assign the template to the current page.
5. Select Insert – Header and select the new page template from the list.
6. Enter the desired text for the header.
7. Apply the command Insert – Manual page break.
8. In the area Type select the option Page break and then in the field Template the template "Standard".
9. Redo the steps 2 to 6 to create a second, user defined page template with a different header.

On the tab **Management** select the next template, which is already defined to follow after the paragraph.

If you want to **assign a formatting template**, you can mark the text or place the cursor as desired and double-click the template in the toolbox. Alternatively you can use the serial mode (bucket symbol), mark a template in the list and click on the objects in the text which shall be formatted with the template. In the window Formatting templates at the bottom there is a **drop down menu**, where you can select, **which templates are listed**: all templates, all templates currently in use or all user defined templates. The **option Hierarchy** shows, which template is based on which other template.

If you have defined your page templates as desired, you can save your document with all formatting templates as a **document template** under a name and from then on create new Writer documents based on this document template.

Linking several documents e. g. several chapter files to a book is done with the function **Global document**. It is useful, if the global document and the subdocuments are based on the same document template. Please read the online help for more details.

A second, very useful help for editing is the **Navigator**. You can open it with **F5**. It shows your **document structure** and shows all **headings, tables, text frames, graphics, OLE objects, text markers, areas, hyperlinks, references, lists, notes and drawing objects**. If you have used these objects in your current document, there is a small plus in front of the object type name. Click on the plus symbol to see a list with the objects contained in your current document. The plus symbol changes to a minus symbol, which can be used to close this branch of the object tree.

If you click on an object in the navigator, the editing cursor jumps to the appertaining location in your document.

There are quite interesting AutoCorrection options in Writer: At the very end of the list you can connect paragraphs which consist of only one line. In the menus **AutoFormat** and **AutoCorrection** Writer distinguishes between three modes of operation during editing. The first two modes are Apply and Apply and mark changes. The third mode reminds of the Mark changes function in Word.

If after you have inserted a **figure** you want to enter a **figure subheading**, click the figure with the right mouse button and insert a label, category Illustration. The two objects are automatically kept together during page break.

You can influence the numbering under Special – Options. On the left side you have to select OpenOfficeOrg Writer and Auto-Labeling. On the right side you have to tick OpenOfficeOrg Writer Image and to change the settings. If the label shall be "Figure" instead of "Illustration", you can overwrite the existing text.

Decorative elements like colored bars, bullets and backgrounds are available under Special – Gallery. Just draw the desired decorative element with the mouse from the gallery to your document.

Writer has a different philosophy than Word for **inserting cross-references. At first** you have to define a **named target** with Insert – Cross-reference – Define reference, then you can refer to it with Insert – Cross-reference – Insert reference. An interesting option is **Above/Below**. Writer automatically sets the option to "above" or "below", depending on where the reference target is, i. e. you can enter a cross-reference "see above/below" and when you edit your text, it is automatically updated. **Inserting hyperlinks** is similar as in Word. If the website shall be opened in a new window, you have to select the option "_blank" under **Frame**. If the link shall appear as a **button**, you have to select "Button" instead of "Text" under the option Form.

When you insert a **hard page break**, you have to define, which page template the new page shall use and which page number it shall have. Inserting **footnotes** is similar as in Word.

Creating a **table of contents and list of figures** is as easy as it is in Word. Please select carefully, whether you want to allow manual editing after the creation of the lists to optimize the line and page break or not. Please read the online help for details.

One strength of Open Office Writer is the management of sources of reference in a **literature data base**. Then you can insert cross-references to literature sources in an elegant way and create a list of references in different output formats (for a journal, the institute or a diploma thesis). Please read the online help for details.

The **data exchange with Word** does not always work smoothly. If you save documents in Open Office Writer in Word format and open them in Word and you do not get the desired result, please try the data exchange via the RTF format or copy the text to the clipboard and insert it into a new, empty document created by Word. In most cases you have to insert the figures again.

Exporting your file **in PDF format** works without problems. Using the option Tagged PDF results in preview pages and a table of contents on the left side. The table of contents contains one entry for each document part heading marked with the appertaining formatting templates in the text. The entries in the table of contents are clickable links. **Saving your document in HTML format** is not very comfortable, since the links cannot be exported.

3.8 Creating slides with presentation graphics programs

Slide presentations are the most common means to visualize oral presentations, see chapter 5, but slide presentations are also used as an alternative to text files to present technical information. Such presentations often occur in the following situations:

- A project leader summarizes the current state of his project in a PowerPoint file and sends it instead of a **project report** with long text **via e-mail**.
- A presentation serves as **product information** for customers, i. e. for promotion.
- The presentation is created for self-study/e-learning, i. e. for **instructional purposes**.

Since the viewing situation when looking at such a presentation for self-study is completely different from looking at presentation slides during an oral presentation, different rules apply for the slide design.

In this section you will find hints for the design of slide presentations for self-study and smaller meetings. These hints refer to PowerPoint 97 or 2000/2003 resp. and OpenOffice Impress.

This book can only show the basics. You can find more detailed hints in the online help, the publications listed in the source of references and the corresponding literature.

3.8.1 Slide creation with PowerPoint

For Windows users Microsoft Office is by far the most-used program package for creating texts, tables and slide presentations.

☞ *If you already use Microsoft Office 2007 and you do not find the functions any more, look here:*
http://office.microsoft.com/en-us/word/HA100744321033.aspx (Word)
http://office.microsoft.com/en-us/help/HA101490761033.aspx (PowerPoint)

Basic layout of the slides
At first you should define the **basic settings for the title and content slides**. In PowerPoint open the menu **View – Master – Slide Master**. Here you can influence the layout of the "Bullet" or "Text" object, which is contained in the preset slide layouts Bullet List, Two-column Text, Text and ClipArt etc. On the slide master you can define

- font type, color and size (Format – Font),
- language (Tools – Language),
- tabs,

- shape and color of the bullets for bullet lists (Format – Bullets and Numbering) and
- line spacing (Format – Line Spacing).

The design of the slide background can be defined in the menu Format – Background. There you can select background color, graphics, textures, color blends and evtl. insert the logo of your company or institute.

In PowerPoint 97 the settings for the title master are defined in menu View – Master – Title Master, i. e. you have to redo the settings there.

If you notice later during slide creation, that something basic is still wrong in the bullet list (text) fields or other systematically repeated elements on your slides, you should change that preferably on your slide master(s).

Font type and size

It has proven to be well-readable and attractive for your readers to use sans-serife fonts like Arial for presentation slides. The font size should not be smaller than 18 pt for beamer presentations, a larger size would be better. For self-study slides the minimum font size should be 12 pt.

Usage of color

Colors can be applied as text colors, background colors and object colors in figures. If you apply them with care, they can **help to mark, emphasize, structure, order and distinguish details**. Colors control reactions, stimulate imaginations, rise **emotions** and bring up memories and associations. Mark the same things with the same colors and use colors **consistently**! If you use color sparingly, it serves as eye-catcher and helps to emphasize important items.

However, color reduces the contrast between foreground and background and reduces the recognizability. You can recognize best black on white. If text is written in dark blue, dark green and dark red on white the letters must be 1.5 times larger and lines accordingly thicker to achieve the same readability. Too many colors and an unclear usage of colors support a sensory overload and tend to burden the overall impression of the display.

Readability test: Display a critical slide on the screen in slide show mode. Measure the width of your screen with a ruler or folding rule, multiply the value by 6 or even by 8 and position your eyes in this distance from the screen. Or print out the critical slide on DIN A4 paper, lay the paper on the floor and look at it while standing. If you can read everything, it is OK.

In Word you can enter a dash with Ctrl + Minus in the numeric keypad. In PowerPoint you can use the key combination Alt + 0150 in the numeric keypad.

To enter text you should preferably use the **standard text object(s) with the bullet lists from the standard slide layouts in PowerPoint**. The indention depth of your bullet lists can be changed with the buttons ⇨ and ⇦ in the tool bar Format. Now you can **pass on the texts on your slides to Word** with the function File – Send to – Microsoft Word and create a handout or script based on the slide texts. Another useful function, which does only work with the predefined slide objects is the **quick entry of text slides in the outline view.** Press the Enter key to create a new slide. If you type in text now this is the slide title. Press Enter again. A new slide occurs. Mark this slide and indent it with the right arrow button. Now you can enter the text for your first new slide. Sub items are created, if you mark and indent them with the right arrow key. **Please note:** All features described in this paragraph work only for the text fields with the bullet lists on the standard slide layouts, they do not work for the text fields which you can insert via the Drawing tool bar.

Inserting text fields: If you click into a text field, it is marked and surrounded by a hatched border. Click on the border and the hatching is changed to many small dots. Now you can position your text field with the arrow keys. A more precise positioning is possible, if you press the Ctrl key, keep it pressed and then use the arrow keys. If the text field is marked with the many small dots, you can copy a text field on a slide with Ctrl-C and insert it at the same position on the next slide with Ctrl-V. Moreover you can delete it with the Del key. **Please note:** All features described in this paragraph work for the text fields with the bullet lists on the standard slide layouts as well as for the text fields which you can insert via the Drawing tool bar.

There are many optional ways to **insert figures**, which can be described here only rudimentary. For example, click on a slide with the right mouse key and in the context menu select slide layouts. Choose the layout Title, Text and ClipArt, enter the title, your text on the left side and use the menu Insert – Picture – ClipArt or Insert – Picture – From file to insert a graphic on the right side.

If you want to **insert a screenshot**, insert a new slide e. g. with the standard layout Title, Text and ClipArt, change to the program where you want to take a screen shot and use the key combination Alt + PrtScr to copy the current window to the clipboard. Now you can change back to PowerPoint and select the object area for the ClipArt. If you paste the clipboard to the area now, the contents of the clipboard is adapted to the proportions of the clipart object. That means it is usually distorted very much. It is better to mark the object area for the ClipArt and Delete it with the Del key. Now you can paste the screen shot with Ctrl-V. Click on the screen shot, and scale it as desired. You should touch the screenshot only in the corners (not at the sides) to scale proportionally. You can enter numerical scale factors, if you click on the graphic with the right mouse button, select the menu item Format graphic and change to the tab Size.

If you want to **tailor your graphic**, this can be done with the Tailor tool in the Graphics toolbar. This lets disappear areas of the image (image margins are cut off), but they are still existing. As just described for screenshots you can **change the image size** in menu Format – Graphic, tab Size.

LIVERPOOL JOHN MOORES UNIVERSITY
LEARNING SERVICES

3.8.2 Slide creation with Open Office Impress

There is the program Open Office Impress for **Linux** and Windows to create slide presentations. This program is very similar to Microsoft PowerPoint. In the following you will find peculiarities during the creation of a simple slide show with Impress which stand out to an experienced PowerPoint user.

After you have started the program there is an assistant who guides you. Open an **empty presentation**. In the function **Choose slide background** you should keep the settings "<Original>" and output to Screen and in **Slide transition** you should select "No effect". A new empty presentation is opened. On the left side you see thumbnails of the slides, in the middle is the working area and on the right side there is the task area with the menu areas Master pages, Layouts, User-defined animation and Slide transition. At the beginning the area Layouts is opened.

You can insert new slides as in PowerPoint in the menu Insert – Page, unfortunately there is no short key command for this. Working in the **outline view** is **nearly identical** as in PowerPoint.

New slides are not inserted by default in the layout Title and Text, but in the **layout last used**. The preset layout is "Empty Slide". To assign a new layout to the current slide, you click on the desired layout in the task area. Therefore you should go to the left and select a slide by clicking on the appertaining thumbnail and then select a layout by clicking it in the task area on the right side. This layout is applied to the current slide in the working area.

The **font names** are different. Arial corresponds to Helvetica, Times New Roman corresponds to New Century Schoolbook and Webdings is similar to Zapf Dingbats. By trend the standard settings result in smooth and more serious presentations.

Editing text in the preset **layout area** (here: in the bullet list field) works as usual, but by default there are nine indention levels in the bullet lists.

If you want to have the (rough) structure of your presentation in a text field at the left side on every slide for **achieving transparency,** you should create a text field on the first slide after the slide with the presentation title and the structure slide. However, this text field cannot be selected as graphics object, copied on one slide and pasted to another slide at exactly the same position. Therefore you should create a slide with the text field containing your structure and duplicate this slide (menu Insert – Duplicate Page).

The **formula editor** (Insert – Object – Formula) is strange at first sight. Use the formula assistant to insert the commands and enter your desired numbers and formula symbols at the bottom area of the screen.

If you use Insert – Picture, there are only "From File..." and "Scan", the **Clip-Arts** known in Microsoft Office are **missing**. The functions Print, Save and Export to PDF work well.

If you want to use **another master page**, on the right side in the task area you have to open the menu for the master pages. The master page which is currently in use is displayed with a thicker border. If you select another master page, it will be applied to the whole slide presentation. To look at the **slide master** and **change the basic settings** for font type and size, bullets, indention values etc., open the menu "View – Master – Slide Master".

To assign a **user-defined animation**, you have to select an object which you want to animate in the working area, e. g. a text field, the whole text should be marked. On the right side in the task area in the menu User-defined Animations click on Add and select an animation, e. g. Appear. By default the whole text is shown at the same time. To let the menu items appear one after the other, click the objects one after the other on the right side in the task area in the list and change the option Start from "With Previous" to "On Click". In the list there are symbols, which are similar to those in PowerPoint. Assigning **slide transitions** works in the same way as in PowerPoint.

If you have created **graphic objects** with the functions in the **Drawing toolbar**, you can modify their look in menu Format in the submenus Area and Line. You can **duplicate graphic and text objects as well as layout areas** in the menu Edit – Duplicate.

3.9 Completion of the Technical Report

The phase "completion of the Technical Report" consists of the tasks proof-reading, entering the corrections, creating the master printouts, end-check as well as copying, binding and distributing the report or creating a PDF file from it and publish it in a data network. No later than directly before the last proof-reading you should talk with your copy shop or print shop, whether you shall provide **digital data or printed master copies**. For digital printing it is often necessary, that you use a special printer driver. This often influences the line and page break.

3.9.1 The report checklist assures quality and completeness

In this phase there is mostly a lot of time-pressure. To find as many errors as possible despite the time-pressure, we have collected our experiences in the following report checklist, **Checklist 3-14**, which shall serve you as a companion during all tasks in the phase "completion of the Technical Report". If you check all points in your Technical Report which are contained in the list and correct them as necessary, there is a high probability that the most frequent mistakes are eliminated. Therefore we recommend, that you check your Technical Report with this report checklist. The quality which can be assured with this approach can hardly be achieved without using the report checklist.

Checklist 3-14 Report checklist

Table of Contents:
- Heading "Contents" (according to ISO 2145)?
- Are there page numbers? is only the first page of every document part listed?
- Are page numbers aligned right-justified?
- Document part number, heading and page number must be identical in the text and in the table of contents!
- Has the automatically created table of contents been updated?

Contents:
- The names of sub functions must be identical in the verbal assessment, in the morphological box and in all other descriptions in the Technical Report regarding number, name and order.
- Reduce sentences in your mind to check, whether they are logic, i e. leave out sub clauses and ellipses. Is there a complete sentence left?

Spelling and layout:
- Is the text left- and right-justified?
- Is automatic hyphenation switched on?
- Is the sign for a range of numerical values either "to" or "–"?
- No space character before full stop, colon, comma, semicolon, exclamation mark.
- Always use a tab between bullet and list item.
- Are bullets used consistently on different hierarchy levels in lists?
- There is always a space character between physical value and physical unit.
- Before and after each mathematical operator (+, –, x, :, <, >, = etc.) there must be one space character. The signs + and - are placed before the numerical value without space character.
- All types of brackets, i. e. (...), [...], /.../, {...} and <...> enclose the text directly without space characters. The same is true for quotes "" and ''.

Figures and tables:
- Figures have a figure subheading, tables a table heading.
- Avoid too thin lines (line thickness ¼ pt) because of problems when copying.
- Figure and table titles always begin with a capital letter.
- Adjectives in tables and lists of requirements should consistently begin with small letters (possible exceptions: table header and introductory column.

Technical drawings:
- Is the title block filled in completely?
- Single part drawings: Can the part be manufactured? Are all measures available which are needed for manufacturing?
- Assembly drawings: Can all parts be mounted? Are all required assembly dimensions given?

Literature work:
- Bibliographical data for cited literature and information where to find the literature in the library should be noted on the first page of the copied sheets directly after you have copied the pages.
- Check these data again before you give back the literature to the library.
- Check bibliographical data in the list of references for completeness.

Line breaks:
- Are the results of the automatic hyphenation correct?
- If there are too wide gaps between words (when text is left- and right-justified) or if the line width differs too much (when text is left-justified) insert hyphenation proposals (Ctrl-Minus).
- If possible, after hyphenation there should not be one syllable alone in a line.

Page breaks:
- Document part heading alone at the bottom of a page and appertaining text on the next page is a no-go!
- One line standing alone at the end or beginning of a page should be avoided.
- Landscape format must be readable, if you turn the page by 90° clockwise (binding margin is at the top).

Miscellaneous:
- Cross references in the text referring to document part numbers, figure and table numbers, page numbers, document part headings etc. should be checked for correct numbers and compared with the table of contents, list of figures, list of tables, etc.
- Folded tables and figures (swing-outs) should be folded so that they are not cut into pieces, when the report is cut after binding.

Completion of the Technical Report:
- Plan sufficient time for proof-reading and end-check!
- If you use formatting templates and automatic creation of lists and cross-references, all fields must be updated (press F9 key).
- When proof-reading, check the text of cross-references to figures and tables, whether you refer to the right figure or table resp.
- Pages, which must be copied with photo key, should be processed separately.
- Figures, which need to be glued in, must be cut so that they have straight and rectangular edges. They should be glued onto the copy original pages with drawing board and ruler.
- Always use a new binder.

3.9.2 Proof-reading and text correction according to ISO 5776

During the final proof-reading (see the next figure) all remaining errors in the Technical Report should be found. The following error types are nearly always still existing in the Technical Report: wrong wording or wrong vocabulary (if the report is written in a foreign language), spelling errors, wrong commas, grammar mistakes, bad writing style, repetition of contents, errors in the logical sequence of thoughts, layout errors and miscellaneous errors (readability, bad briefing of the typist).

If you can concentrate well, search all errors in one proof-reading cycle. Otherwise group the types of errors, you want to concentrate on, into groups and **search the errors in several proof-reading cycles.** You can find all error types listed above far better, if you do the proof-reading on paper printouts rather than on the screen. **On paper** the contours of the letters have more contrast and sharpness. In addition, on a sheet of paper you can look at far more information in a context as on a screen and turning the pages back and forth is much easier. This facilitates looking for repetitions of contents a lot. Enter the corrections, which you marked on the paper very soon into the appropriate text files!

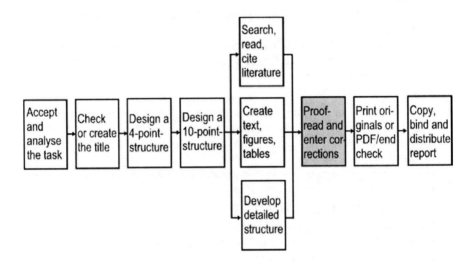

While entering the corrections and reading the text on the screen, you will often find additional **open problems.** Try to become used to standardized personal notes and symbols for those things, which you have to check later, before the final printout is made, but which would disturb your workflow too much at the moment. For example, use a **search marker** like "###", which is entered into the file at the desired position, where you want to jump to directly. You can add abbreviations to this search marker like "###Sp" = spelling/dictionary, "###Lit" = literature citation or check again bibliographical data, "###Un" = meaning is unknown (look-up the details in a textbook or encyclopedia), "###iS" = introductory or connecting sentence etc.

☞ *Note all remaining necessary corrections and work steps in your to-do-list immediately, so that you do not forget them.*

When proof-reading you also **check the layout**. Figures and tables may not be split apart as long as they fit on one page. The figure subheadings and table headings must appear on the same page as the figure or table. If paragraphs must be split by a page break, no single line should remain on the old or flow to the new page. Also forbidden is, that the document part heading stands at the bottom of the old page and the first paragraph of the text starts on the new page.

Since the page numbers may change until you entered the very last change, the page numbers are the last thing you enter into the manually created table of contents and the automatically created table of contents must be updated again. This is also true for the list of figures, list of tables etc. Therefore the master copies of the table of contents and the other lists are the last master copies that you print out. Before you start copying your Technical Report you should check again all cross-references to other document parts and compare all page numbers of the document part headings in all previously printed master copy pages with the entries in the table of contents.

Corrections should always be entered in red, e. g. with a thin felt pen or ball pen. This is important, because you can then see your own corrections far better during entering them into your computer files as compared with corrections marked with pencil or blue or black ball pen. If you have entered a correction, you should move a ruler down across the page until you reach the next correction symbol. You can as well tick or cross out the correction symbols with a pen or pencil having a different color.

Correction symbols are especially important, if the author of the Technical Report enters the corrections on the proof-printout and someone else enters them into the computer files. You can find the standardized correction symbols in ISO 5776. We recommend a meaningful selection of correction symbols with some simplifications, **Checklist 3-15**.

Checklist 3-15   Simplified system of correction symbols according to ISO 5776

Each correction is
a)   marked in the current line with a **short vertical bar at the left margin** and
b)   marked with a **correction symbol in the text**.
c)   The **correction symbol is repeated at the right margin and** on the right side of it the **correction is noted**.
d)   The repetition of the correction symbol is not required, if it is "self-explaining" like the symbols for insert/delete space character, insert/continue paragraph etc.

| **Correction markings in the text** | **Corrections** |

Letters which must be crossed out are marked like this here| The
letter stands for the Latin word "deleatur" = "it shall be deleted".
The horizontal line with vertical bars (├───┤) is not only used for
deleting text, but also for replacements and for the exchange of
~~specialised~~|letters.

Text areas running across several lines are deleted|~~with a vertical
bar at the beginning and end and a letter z in between~~|.

If only one letter is to be deleted or wrong, then it/corrected as
shown. If the⌐ are several corrections of this typ⌐ in one li⌐, the
correction characters are varied. Each correction character appears
only once in the same line.

The diagonal line with hook(s) crosses out the wrong character, at
the margin there is a replacement for the character. Wrong punctuation should only be underlined in the text; so that you can see
better, what stood there before.

For a different character formatting the **relevant** word or the text
passage is underlined with a wavy line and the desired or wrong
formatting is noted in the margin, e. g.: i = italic, b = bold,
u = underlined etc.

Sometimes a ~~really~~ undesired correction must be canceled.

If larger amounts of text must be inserted as correction, for which
the space in the right margin is not sufficient, use graphic sym-
bols. These symbols are repeated at a location, where there is
enough space for the missing text (page margin, reverse side).
You can use the following graphic symbols: ◯,⊗,◻,⊠,△,▽etc.
If a character must be replaced by a space character, use the
symbol at the right. Example: DIN/A3.
If a space character is missing, this marking is inserted.
If there is a space character too much, it must be deleted.
If a blank line is missing, this must also be marked.
So, if a new paragraph starts here, the blank line is missing. This
sentence is he last sentence of a paragraph.

The line of thinking is now continued and the paragraph shall be
omitted. This is marked with the upwards directed double curve.
It may also happen, that a paragraph must be inserted into a text.
This is also explicitly marked.|After introducing and explaining
the correction symbols we want to give you some hints for the final printout now.

Sometimes the beginning of a line is not where it should be. This
can be corrected in a self-explaining way by marking
the word with the wrong indent and the left margin with vertical
lines and drawing an offset arrow.

This method is used for more and less indenting (moving the word to the left and to the right).

You can see the correction symbol exchange to letters or words which are in wrong order ~~without mnemonic~~ as shown in the current the sentenced.

☞ *The symbol to insert a blank line ">" can be completed with "⌐S>" for an introductory/connecting sentence, with "P6>" for a page break, with "+>" for more vertical space and with "–>" for less vertical space.*

You can learn and immediately use this simplified system of correction symbols by simply reading the checklist above, since all symbols are logic in themselves and can be deducted again easily. These correction symbols have proven its value in practice very well.

Beside these correction symbols you can use correction markings which are to be worked on later. During the phase "proofreading and correcting" you will probably work on the "###xxx" markings in the text with appropriate corrections and delete the markings as well as rework the text style. But even when you enter the corrections, you may still decide to leave some items unsolved, e. g. because special textbooks or encyclopediae can only be consulted during the next visit to the library. You should structure this phase of **inserting the corrections** as follows:

- In the printout **at the left margin mark all lines containing the "###Sp", "###Lit", "###Un" und "###iS"** with a vertical line with a well visible (colored) pencil or text marker. In addition, write appropriate notes what to do into the margin, e. g. in which book you want to read again the marked issue. With these well visible (colored) vertical lines you can find these editing notes quickly in the end phase which is characterized by time pressure.
- **Cross out all vertical lines in the left margin, as soon as you have entered the correction into your file**, if possible with a pencil in a different color like green.
- **Still necessary work steps** should be ordered by category, **copied into a file as to-do-list ordered by category** and printed out, resolved items in the list should be crossed out, this should be updated in your file etc.
- Work on the last items of your to-do-list or speak with your customer or supervisor to make sure, that the last changes are not necessary any more.

If possible, let others also read your Technical Report. After several weeks or after frequent reading of your own texts you will become "routine-blinded" against your own formulations. The following persons are candidates for proofreading: father/mother, partner, boyfriend/girlfriend, fellow students, supervisor, professor, customer etc. They should have at least one of the two following qualifications:

- expert for the project/*specialist* and
- expert for spelling or foreign languages/*generalist.*

Evtl. a different person should do the proof-reading for the other area.

☞ *If you have entered the corrections, tidy up the old pages. Too many text versions tend to get mixed up. You should only keep the latest version with corrections and throw it away right after you have created the final printout.*

If you have **entered the last correction**, no open correction markings are left and when you are about to do the final printout, **check the line break** again on the screen! It often happens, that due to text changes hyphens appear in the middle of the text or soft hyphens are forgotten and the left and right-justification creates undesired gaps between the words. Define, whether your word processor shall do automatic or semi-automatic hyphenation (hyphenation proposals). Automatic hyphenation sometimes leads to mistakes.

Then you should use the page preview with a zoom factor, so that you can see the whole page to check again the **page break and layout**. Figures and tables and their appertaining headings may not be cut at wrong places by the page breaks. The distances above and below formulas, examples, figure subheadings, table headings etc. should be consistent. The headings should be near the object they refer to (e. g. table, figure). The rest of the text has a considerably larger distance. If you have optimized the line and page break now, the text is ready for the final printout.

3.9.3 Creating and printing the copy originals and end check

The text, tables and figures are ready now. You have proofread the drafts and evtl. gave them to other people for proofreading. The corrections are inserted and the final printout could start now.

But before you start with that, you should call your **copy-shop** and **announce your copy order**. Tell them, how many copies are to be made and when you will bring the copy originals. Besides you should ask, how long it lasts to create the bindings (cold adhesive bindings need ≥ 1 day to dry). If you have arranged the copying of larger works or larger print runs with the copy-shop, you can start to create the copy originals. In the network plan this is the second last phase of creating the Technical Report.

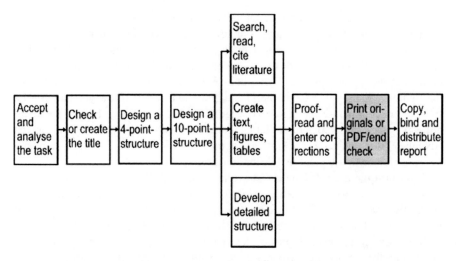

While you can use nearly any paper for drafts, you want to get as **high-class copy originals** as possible. High-class means, that the copy originals should be **rich in contrast**. That means at first, that the **printing paper should be pure white**. That is also valid for figures, which are copied from another source and integrated into your report. Also, the paper shall not shine through. Therefore you should use paper with a mass of unit area of 80 g/m^2.

For **ink jet printers** you should use as **smooth paper** as possible, evtl. even special paper. If you want to integrate a few copied pages into your report or glue-in copied figures, you should use the same paper for these purposes. That means, that you should fill the copier with your own paper, if necessary. **Do not use** pure white paper in your own printer and **ecological or recycling paper in the copier**. This would result in an avoidable inconsistent impression. Naturally speaking, color copies may be on different paper, because you normally cannot influence the paper quality for these copiers.

Manually drawn figures (freehand drawings or drawn with rulers and stencils) appear especially pretty and rich of contrast, if you **draw them larger** and **reduce them with the copier**. Small irregularities and blurs disappear. The lines appear much sharper.

The importance of manual drawing declines, but compared with using graphics and CAD programs it still has a certain significance. The phases "creative sketching" and "drawing ready for printing" seamlessly merge when drawing with a drop action pencil and you can rub out errors easily. The copier gets out additional contrast, if you use pure white paper and a high contrast setting. Especially, if you run out of time towards the end of your project, you should seriously check this alternative. Also, in our experience you can concentrate on drawing with few errors after many hours and late at night better, if you are working manually. With the computer you have quickly pressed a button, you did not want to!

If copied or cut-out figures or tables shall be glued in, always use **drawing board and ruler**. Clamp the paper and adjust the paper edge. Then lay the figure loose on the paper. Decide, whether the figures shall be arranged left-justified or centered or with a regular offset of a few centimeters from the left edge. Roughly remember the figure position. Then put as few glue as possible on the rear side of the figure (e. g. only into the corners) and glue the figure in. Now you can use the ruler to check whether the figure is straight. If not, move it carefully. **Twist-up adhesive stick (glue stick)** is reliable. Compared with all-purpose glue it has the advantage, that you can use a knife to get off the figures again, To adjust them at a new position or to use them in a different context.

The **montage glue "Fixogum"** has the advantage, that the figures can be removed (stripped off) very easily. You can even get off newspaper paper. However, this glue has the disadvantage, that after four or five years all figures that you have glued in become loose and that the paper gets yellow at the glue spots!

If you use **transparency paper**, the glue shines through (evtl. copy the image or drawing from transparency paper to pure white paper and glue that in).

When **copying**, the **edges** of the glued in figures might be **visible** on the copies. Then you have two options: Change the contrast at the copier or – if the copies may not become brighter – cover the edges on the copy original with correction fluid. You can also make a good, dark copy and cover the undesired edges with correction fluid there. Then this processed copy becomes the new copy original.

For all labels you want to write by hand and colored accentuations of different information (e. g. distinguishing the variants in the morphological box, colored underlining etc.) there are some rules for the selection of drawing pencils, if you do not use the color ink jet printer.

If you draw with **felt pens**, you can usually see the color also on the rear side. It is better to use text marker, pencil, colored pencil, ball pen and colored india ink. However, ball pens often "spit" (especially before the mina becomes empty) and create thick blurs at the beginning and end of the line, which dry out very slowly and which are therefore smeared very easily. Refilled text markers also spit very often! India ink pencils create the most accurate results; but since you have to wait for any line that the ink is dried, they take a lot of patience.

The **colors yellow, light brown, light green, light blue** etc. are **hard to copy with black-and-white copiers**. In the copy they are either ignored or displayed as very light gray. The colors dark blue, dark green, red etc. can be copied well. They are displayed as dark gray to black.

If **only the master copy of the Technical Report** shall get **colored accentuations**, then you should use strong, dark colors and different line styles on your copy originals. On all other copies but the master copy the different black line styles are clearly distinguishable.

If you want to **glue in labels into figures**, you should leave at least two spare lines between the different labels on your paper printout. Then you cut out the labels and glue them into the figures. The procedure is the same as described above for gluing in figures. To glue in labels you should use Twist-up adhesive stick (glue stick). Liquid glues flow out at the sides of the narrow paper stripes and excessive glue, which has not been removed in time would glue together the copy original pages or would be visible on the copies.

Figures and tables can be **larger than DIN A4**. This is called "swing-out". Prepare a nearly empty page as copy original for such a picture or table ("carrier page"). On this **carrier page** there are only the normal header with a running page number, evtl. a footer and the figure subheading or table heading. The figure or table is created in an independent file and printed out or copied (evtl. on a color copier) or plotted as often as the number of copies of the final bound Technical Report.

After copying the Technical Report including the carrier page the **swing-out is glued onto the carrier page**. All **left bending edges** must be clearly **out of the binding range**. The **right and lower bending edges** must have a **distance of five (better ten) millimeters from the paper edge**, since for some binding types after binding the document is cut with the hydraulic shear at the upper, right and lower edge.

If you have used a technical part or assembly from a **manufacturer brochure** in your design, it is better for the understanding of your readers to include the used brochure into **an appendix** of your Technical Report. If you do that, another step is recommended very much. Help your readers to quickly find the **used formula, selected measure etc.** by **accentuating** that in the brochure in the appendix **with a text marker**.

After all figures and tables are glued in, all manually created labels are written and all correction steps are done the end check can follow:

Are the pages in the **right order**? Are they upright (page headers at the top)? **Pages in landscape format** lie correctly, if the upper page margin which is to be bound is on the left side in the pile of sheets and the pages can be read from the right. Have **outdated printouts** slipped-in? Go through the **report checklist, Checklist 3-14**.

If everything is checked – and also the page numbers in the table of contents have been checked with the copy originals –, you can **go to the copy-shop**.

3.9.4 Exporting the Technical Report to HTML or PDF for publication

If you want to publish your Technical Report in a data network, i. e. in an intranet or in the internet, there are two popular formats available: HTML and PDF. The PDF format (Portable Document Format) was defined by Adobe. To create PDF files you need the commercial software Adobe Acrobat, (the free-of-charge Adobe Acrobat Reader is not sufficient for that) or another PDF writer software or a printer driver that creates PDF files. Below it is described how you can save your Technical Report as PDF file.

PDF is a page description language. Every reader sees your Technical Report on the screen in exactly the same way as you have designed it on your computer. The line and page break is kept. That is the main advantage of PDF.

On the other hand, if you would publish a Word file in an intranet, the line and page break usually gets lost, because the currently installed printer driver on the other computer is used. For the correct display of the Technical Report the reader also needs the same fonts as the author.

HTML (Hypertext Markup Language) is the page description language used in the internet and in an intranet. Every reader sees your Technical Report as the currently used browser interprets the commands. The line and page break, you have intended, gets lost. That is the main disadvantage of HTML. The main advantage is, that HTML files are smaller than PDF files. Therefore they can be displayed much faster.

You should also know the principle of linking pages in HTML and PDF for working with Word, PowerPoint and Excel (and similar office programs), because these office programs offer functions to link office documents among each other and with external files since quite a while. The following sections describe the most important functions.

Automatically created links (hyperlinks):
If you use the functions described in 3.7.4, Word creates the features needed for the publication on paper (entries in lists, tables and indexes, labels, cross-references) and in addition it automatically creates the links needed for online publication in the background. You can see this e. g., because all cross-references in the document are clickable and have the paragraph format Hyperlink. Also internet addresses are automatically converted to hyperlinks. The prerequisite is that the address either starts with "www." or with "http://". The automatic conversion of internet addresses to clickable hyperlinks works in Excel, PowerPoint, Works etc., too. Also, the functions of Open Office Writer and Open Office Impress described in 3.7.5 and 3.8.2 create clickable hyperlinks from the internet addresses.

Manually created links:
In Word you need the functions "Insert – Text marker" and "Insert – Hyperlink" to link pages. A **hyperlink** is a connection, a reference or a goto *command*. You can link each clickable object including images. If you click on the hyperlink, you go to another position within the same file or a program is started and the file or internet address named in the hyperlink is displayed. The other program can be PowerPoint or Real Player, for example. It depends on the file extension, which other program is started (see Windows Explorer: Special – Folder options – File types). A **text marker** is the *address* of a goto command within a file. Try it out to define text markers and hyperlinks within a Word file. If necessary, refer to the Word online help.

If you want to manually create links in Excel and PowerPoint, you can use the function "Insert – Hyperlink" as in Word.

Following a link:

If you move the mouse cursor over a link in a Word, PowerPoint or Excel file, the cursor changes. The cursor shape becomes a hand pointing upwards. In addition, an information text in a yellow box appears. It either tells the link address – then you can click it directly – or it asks you to press the Ctrl key and the left mouse button at the same time to go to the link address (depending on the program version). The office programs show these messages for automatically created as well as for manually created hyperlinks.

By the way, the cursor shape changes in the same way, if you move the cursor over a link in an HTML or PDF file, but the information text in the yellow box appears only in the office programs.

Export from Word to HTML and PDF:

The references, which you have manually defined in a Word file and the references Word automatically creates remain 1 : 1 unchanged, if you save the file with "File – Save as Webpage" in HTML format or with Adobe Acrobat in PDF format. When exporting to HTML, each word file results in one HTML file. You have to manually insert links between the files (e. g. one file for each chapter) or manually create an HTML ToC file.

When exporting to PDF, please keep in mind the following: Depending on the program version of Adobe Acrobat it behaved sometimes so that the displayed PDF file did not have a file name yet. If you use this version, you have to get used to saving the PDF file directly after its creation. Otherwise it gets lost, when you close the Acrobat Reader or Acrobat window. To control the access permissions to the PDF file, please refer to the menu "Acrobat – Change conversion settings – Security". If the created PDF file shall contain your document part headings as a tree structure to navigate at the left side of the file, you have to open the menu "Acrobat – Change conversion settings – Bookmarks" and select, which headings and other paragraph format templates shall be used for the creation of bookmarks.

3.9.5 Copying, binding or stapling the Technical Report and distribution

Copying, binding or stapling and distribution of the Technical Report is the last step in the network plan for the creation of Technical Reports.

If you want to distribute or publish your Technical Report online, you have to save the final version of your data again in PDF (or HTML) format, see 3.9.4, and send the files to your webmaster.

If you want to distribute or publish your Technical Report in hardcopy form, it has to be copied or printed first. In most cases it will be copied, because the print run is small. All copies shall be made on the same copying paper. This is also true, if several authors work together to create the Technical Report. One single exception is admitted: There are color copies in your Technical Report. In this case there is usually different paper required because of the different copying process for color copies and therefore it is admitted.

After copying the report is bound. This makes a readable document of it, that – depending on the number of pages and binding type – can have the character of a script, booklet or book.

In the copy-shop you should stay a while in case there are open questions. There might be questions like: Shall we align the cover sheet a little different or enlarge it a little? Which plastic spiral (comb) or binding wire or fabric tape is desired? Which cardboard with which surface structure and color shall we use? etc.

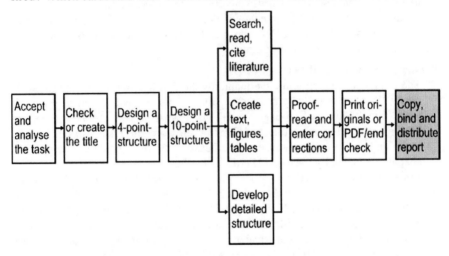

The following stapling and binding types are available:

- staple/paper-clip
- saddle-stitching (= binding type of journals)
- plastic folder
- filing fastener (= plastic stripes with holes and metal tongues)
- spring binder
- spring strip
- file binder
- ring binder
- comb binding (spiral binding with plastic spirals)
- wire-O binding (spiral binding with wire spirals)
- staple binding
- cold adhesive binding
- hot adhesive binding
- hardcover binding

These stapling and binding types differ a lot in price and handling properties. To select the best suited binding for the current purpose, we want to list those points at first, which have to be done before the binding, and then we shortly describe the properties (advantages/disadvantages) of the various bindings. For determining the binding type you should speak with your customer or supervisor.

To be done before the binding:

- If you want to bind swing-outs (= figures, tables or lists, which are wider or larger than DIN A4): are the bends on the right and at the bottom far enough away from the paper edges (≥ 5 mm), so that they do not fall into pieces when the report is sheared? When the report is shaken in the copy-shop they have to make sure that the pages with the swing-outs move into the binding area correctly! This is especially important for adhesive bindings. Problems can be avoided, if you let them bind a DIN A4 carrier page into the report and glue the swing-out onto the carrier page after the binding.
- If you have drawings, tables, figures or lists, which are larger than DIN A4, you have to decide and keep in mind the following points:
    - **Swing-outs to the bottom**, should be folded and glued as follows:

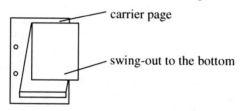

    - Shall copies of drawings be bound into the Technical Report?
      If yes: Original drawings in DIN A2 and DIN A3 can be reduced to
      DIN A4 with the copier without problems and normally bound.
      Larger drawings should be reduced with a copier by maximum two DIN
      steps (e. g. DIN A0 to DIN A2). Then fold the copies to the normal paper
      size DIN A4, see image below.
    - The unfolded original drawings are always handed in in a folder or roll.
    - If there are **DIN A3 drawings in landscape format or swing-outs to the right**, you can decide, whether you want to bind them into the report or combine them evtl. together with drawings of other size(s) in a file folder. If the drawings shall be bound together with the report, fold them according to the following scheme. The security distance assures, that the pages of the Technical Report can easily be turned. If you bind this drawing as swing-out into the report, the security distance helps, that the swing-out is not destroyed during shearing. Then the filing holes are not there and on the left side there must be enough space/white room for the binding.

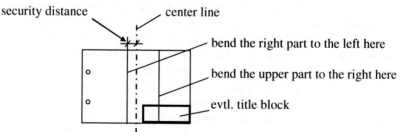

Folding a DIN A3 drawing in landscape format or swing-out to the right

If the drawings shall not be bound into the report, but presented in a file binder, fold them as follows.

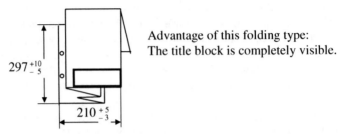

Advantage of this folding type:
The title block is completely visible.

Folding a DIN A1 drawing to DIN A4

The folded drawing (or table or figure) must be unfoldable and refoldable while it is still bound in the file binder. Plastic adhesive **reinforcing rings are recommended**.

- If you want to glue in photos, copied images or images from other printed materials (newspapers, journals, brochures etc.), please do that before binding!
- If you want to use sheet protectors, make sure to use a type that has three closed sides with the open side at the top.
- Sheet protectors cannot be bound into staple binding, the adhesive bindings and hardcover binding, because they would be destroyed during shearing. You can only use sheet protectors together with plastic folder, filing fastener, file binder and ring binder.
- Did you sign yourself the declaration in lieu of an oath after the copying with a document-proof pencil (ballpen or pen)? All copies must be manually signed.
- The binding types staple/paper-clip, filing fastener and spring strip do not provide a container for the sheets of paper. If you have decided to use staple or paper-clip, you can hand in the documents in a sheet-protector (one or two sides open). In case of filing fastener and spring strip the first and last "sheet" can evtl. be a (semi-)transparent plastic or cardboard sheet or a plastic cover. That creates a more tidy impression and protects the paper.
- For the binding types plastic binder, spring binder, file binder and ring binder you should not hand in used containers (with signs of wear).

Often it is useful to bind bulky appendix material like brochures, corporate publications, measuring protocols, program listings, electric plans, wiring diagrams, technical drawings etc. in a separate file binder or to bind them separately. This separate binder belongs to the report. The documents are listed in the list of references. These documents are also added to the table of contents of the Technical Report with the note "(in separate binder)". If you would have to list many documents with that note, it is better to split the table of contents into one for volume 1 and one for volume 2. Both tables of contents are then included into both volumes of the Technical Report. But now the properties of the various binding types are described.

Staple/paper-clip

Advantages: Extremely cheap, after removing the staple or paper-clip the documents can be punched and integrated into the filing system, copying with paper feed is easily possible.

Disadvantages: Only suited for thin documents, to read the inner pages you have to bend the corners. If there are many documents with staples or paper-clips in a folder, the upper left corner becomes very thick.

Saddle-stitching (two staples in the back of a brochure)

Advantages: This binding is used for journals, brochures and booklets of any type, is extremely cheap and requires, that the Technical Report has been reduced during copying (from DIN A4 to DIN A5) and copied double-sided.

Disadvantages: This binding is created with a stapler with longer arm, then the complete report is folded once in the middle. It is possible to process max. 20 sheets (80 pages) with this binding type. If you want to copy the bound report later, you have to manually turn the pages and put the report onto the copier, if the binding shall not be opened.

Plastic folder

Advantages: This binding is very cheap and easy to be done. Sheet protectors can easily be integrated. With this binding type you can bind up to about 100 sheets; if there are more sheets, the report becomes more and more difficult to handle.

Disadvantages: To read the inner pages, you have to bend the pages to the left, because otherwise the report would not stay opened when lying on a table. The metal tongues are often sharp-edged, if the report is read more often, the holes become frayed. Since the binding can be opened easily, copying with paper feed is easily possible, as long as the holes are not too frayed. Therefore this binding type is only suited for your own documents or for unimportant purposes.

Filing fastener

Advantages: Filing fasteners are much cheaper than plastic binders. The handling of the bound documents is comparable with plastic binders.

Disadvantages: The filing fastener does not provide a container for the sheets, it is only possible to file the filing fastener with its cardboard or plastic stripe, onto which the bound sheets are fixed, into a file binder. Filing fasteners are not suited to present your Technical Report in an optically attractive way. Therefore this binding type is only suited for your own documents or for unimportant purposes.

Spring binder

Advantages: Spring binders are available in different variations. The sheets are not punched, but put into a protecting plastic cover and clamped together with a plastic bar with a groove (= spring strip). The spring strips are available in different thicknesses. If the sheets shall not be punched, but still presented in an optically attractive form, this binding type is suited well (like sheet protectors). Further copies can be made easily with paper feed.

Disadvantages: The number of sheets is limited to about 30 to 40 sheets. sheet protectors are hard to clamp, because they are too smooth. The bound report does not stay opened, it shuts itself again.

Spring strip

Advantages: The spring strip is similar to the spring binder, but it is much cheaper, because it does not have a plastic cover.

Disadvantages: The spring strip does not provide a container for the sheets, it only provides a losable connection of the sheets.

File binder

Advantages: You can turn the pages well, the pages remain opened, sheet protectors can be integrated easily, copies can easily be made with paper feed, it is possible to bind comparatively many sheets (in 8 cm file binders you can bind up to 500 sheets without staples and paper-clips).

Disadvantages: File binders usually need much space on the shelves and are rather bulky in opened state. Besides, if you often turn the pages, the holes may become frayed. The only countermeasure is to stick on plastic adhesive reinforcing rings.

File binders are not very attractive for presentation purposes. Therefore select a different binding type for your Technical Report and use a file binder only for an appendix (or several appendices) with brochures, manufacturer catalogues, corporate publications, measuring protocols, program listings, drawings etc. Blueprints, plots or copies of your drawings which are larger than DIN A3 and must be folded to DIN A4.

Originals on transparency paper are *never* folded! They are stored and transported in drawing folders or rolls. During meetings the originals can be fixed at a clamping or magnetic bar at the wall in the meeting room (but even there blueprints, plots or copies are better/safer).

Ring binder

Advantages: Ring binders have for metal rings in Europe and three metal rings in the US which can be opened to insert or take out punched sheets of paper. They provide an optically attractive presentation of smaller Technical Reports. Commonly sheet protectors are used and the sheets are inserted into the protectors (three sides closed, top side open), so that you can read the pages like in a book. Then the pages with odd page numbers should appear on right sides and the pages with even page numbers on left sides.

Disadvantages: Ring binders are usually more expensive than file binders. Inserting the sheets into the sheet protectors takes effort!

Comb binding (spiral binding with plastic spiral)

Advantages: This binding type is relatively compact and cheap. You can put a sticky label with the name of the report (and evtl. author) onto the plastic spiral/comb and the **sheets stay opened when the report is lying on a table**, you can turn the pages of the bound report around nearly completely. With appropriate spirals/combs this binding type can bind up to about 500 sheets.

Disadvantages: The **sheets must be aligned carefully** before they are punched and during punching they must be laid against the stop bar of the machine with care, so that the holes are not too near to the paper edge and tear off. Caution! Do not punch too many sheets at the same time. If you need too much force for punching, the machine might be damaged!

Commonly a **transparent cover** (thicker than overhead slides) is used on top of the title leaf. Caution! This sheet must be **punched separately** in any case. Otherwise the transparent cover and the paper sheets move against each other and the punching is irreparably wrong.

In the copy-shop the sheets are shaken on a machine. There it is also possible to shear the bound Technical Report at the right edge, because the sheets are not lying exactly on top of each other, but they adopt to the curvature of the spiral/comb.

If the **spiral/comb** is **not large enough**, the sheets are bent near the rectangular binding holes and then you cannot turn the pages very easily any more. However, if the spiral/comb is much thicker than the Technical Report, then the report does not stand well on the shelves.

The plastic spirals/combs are often sharp-edged at the corners. **You** should **round off sharp corners with a pair of scissors** before using the comb. If you do not have a machine for creating the binding (i. e. punching and opening and closing the combs), you can only make copies by manually turning the pages and putting the report onto the copier.

Wire-O binding (spiral binding with wire spirals)

Advantages: The wire-O binding is even more compact than the comb binding and about equal in price, but the **sheets do not bend so easily when turning the pages**, because the binding holes are round. With this binding type books can be turned around completely. Therefore this binding type is well-suited for manuals. With this binding type you can bind up to about 350 sheets.

Disadvantages: Putting a sticky label with the name of the report (and evtl. the author) is not possible. The information regarding a transparent plastic cover sheet given under comb binding is valid here as well. For wire-O binding it is also true, that you can hurt yourself (here at the wire) at the upper and lower edge of the report. **Chamfer sharp edges with a file.** Copies can only be made by manually turning the pages and putting the report onto the copier.

Staple binding

Advantages: This binding type is a little cheaper than the spiral bindings. The stack of sheets is shaken with a machine. Then five staples are twinged through the stack and bent on the other side. Then a fabric tape is attached to the binding, which covers the staples, a small stripe on the front and rear side and the spine. Then the book is sheared with the hydraulic shear at the three open edges. **This binding type is extremely stable.** The impression is created, that you are holding a real book in your hands. Due to the stability and the attractive appearance this binding type can be recommended for Technical Reports. For diploma and bachelor/master theses and smaller reports during an engineering study course the staple binding is the most common binding type. With this binding type you can bind up to about 200 sheets.

Disadvantages: The sheets/the book does not stay opened when lying on a table. Therefore in most cases only the front sides of the pages are printed. Copies can only be made by manually turning the pages and putting the report onto the copier.

Cold adhesive binding

Advantages: This binding type is a little cheaper than the spiral bindings. The stack of sheets is shaken with a machine. Then the spine is covered with cold glue, which needs to dry for one to two days. Then a fabric tape is attached to the binding, which covers a small stripe on the front and rear side and the spine. Alternatively, a printed cardboard envelope can be attached, which covers the front side, the spine and the rear side. In this case you can easily write onto the spine whatever you like. Then the book is sheared with the hydraulic shear at the three open edges. The impression is created, that you are holding a real book in your hands. With this binding type you can bind nearly any number of sheets. Due to the possibility, that you can also bind thick books and use a cardboard envelope as well as due to the attractive appearance this binding type can be recommended for Technical Reports.

Disadvantages: Depending on the stiffness of the glue the sheets do not stay opened when the report is lying on a table. Copies can only be made by manually turning the pages and putting the report onto the copier. This binding type is not very stable. It happens often, that a few pages fall out of the glue, especially when the report is copied.

Hot adhesive binding

Advantages: For this binding type the sheets are put into a folder/an envelope, which has a hard stripe of glue in the spine. The folder/envelope is put into a machine which heats up the glue so that it is pasty. Then the folder/envelope with the sheets is taken out and the spine is stuck onto a hard underground so that the paper sinks even deeper into the glue. Now the glue needs to cool for a few minutes to become hard again. This binding type is very attractive for smaller reports, e. g. seminar documentations.

Disadvantages: See disadvantages of cold adhesive binding.

Hardcover binding

Advantages: This binding type is the "most valuable" binding type. It is stable, the sheets of the book remain opened, when the book is lying on a table and they do not fall out of the book. The pages are sheared and the cover can be printed or embossed as you like.

Disadvantages: This binding type is very expensive. Therefore it is usually out of choice for Technical Reports. The only exception might be, that one copy of a bachelor, master, diploma or doctorate thesis is produced with hardcover binding, e. g. with black cover and golden letters.

4 Useful behavior for working on your project and writing the Technical Report

In this chapter we will introduce some useful information and procedures to improve your working effectivity. In this context we would like to recommend reading chapter B "Glossary – terms of printing technology". It does not only contain explanations of technical terms used in this book, but also many definitions of **technical terms in printing technology**, which may come across when you are working together with copy-shops, computer stores, printers, as well as journal and book publishers.

In addition, here are some interesting **links to the internet**, which are probably useful for you when you are writing Technical Reports:

www.leo.org: German-English dictionary and
www.systranbox.com/systran/box: translation of short texts in various directions
http://www.google.de/language_tools?hl=en: translation of short texts in various directions

In the next sections you will get some hints for working together with your supervisor or customer, for working together in a team of authors or a project team, advice for working in a library, for organizing your file structure and creating back-ups of your data, as well as hints for your personal working methodology while writing your Technical Report.

4.1 Working together with the supervisor or customer

You should always try to prepare and to plan ahead the meetings with your supervisor as well as you can. This includes first of all that you will *always* take with you the current version of the **text** <u>and</u> the current **structure** of your Technical Report when you go to a meeting with your supervisor. This helps him/her to improve the supervision and support very much, because he/she can then overview the current stage of your Technical Report or project better and faster.

Before your meeting you should write down **questions** and tick off those questions which have already been treated during your talk. This helps to avoid that you forget something. If you want to take with you many documents and parts for the meeting with your supervisor, a **checklist** might be helpful, where you list all necessary **documents, parts** and **pathnames** to files, which you want to discuss with your supervisor on-screen.

L. Hering, H. Hering, *How to Write Technical Reports*,
DOI 10.1007/978-3-540-69929-3_4, © Springer-Verlag Berlin Heidelberg 2010

When collecting items for the checklist, you should also regard **required projectors and boards** like: beamer for the presentation of files, overhead projector, slide projector, flipchart, poster hanging clips, or magnetic bar to fix plans or drawings etc. You should prepare the meeting in such a way, that all required equipment is available in time. If you want to project pictures make sure, that the meeting room can be shaded. Try out the projection conditions prior to the meeting. Can everybody see the images well?

You should also think about whether a more comprehensive presentation of the results developed so far is necessary. If yes, you should plan this part according to the information provided in chapter 5 "Presenting the Technical Report".

During discussions with your supervisor you should always use the **yes-but-approach**. That means, even, if you have a different opinion, you should agree at first, so that there does not occur any sharpness or denial in the talk. After two or three additional sentences you can carefully change to "no" and wait for the reaction of your supervisor.

☞ *Do not make compromises. Even, if your supervisor does not look into every detail at once, you should stick to this rule: Items, where you are doubtful yourself, are nearly always a source for criticism of the supervisor or customer. This is an often proven fact. Therefore you should not leave formulations you are in doubt about, just because of time-pressure or a research is only possible during your next visit to the library.*

If you are uncertain about a statement, think about the basic rule: "All details and statements which are not written down, cannot be wrong." You should either leave that detail or statement out or research thoroughly, to smooth the uncertainties. If you pass on your Technical Report for **correction**, the following **documents** may be necessary, evtl. as files:

- written task,
- master copies from technical documents,
- technical drawings, on which you build up your work,
- manufacturer brochures,
- current version of drafts (on cardboard or as CAD drawing),
- current version of texts (evtl. with the corrections of the supervisor) etc.

As a precaution you should also take documents with you to the meeting or provide them for correction, which you think you will probably not need. But please *do not present* your Technical Report *in sheet protectors* or provide water-soluble transparency pens.

During your talk **take notes** of hints from your supervisor and **action items still to do**. Think about whether these hints refer to a meaningful specialization of your project, additional literature resources, or new sources of information and take appropriate short notes.

Towards the end of your project you should try to only work on hints of your supervisor, which do not cause too much work. If there are hints causing a high work-load, you should point out the consequences in time and decide for yourself, how much additional work you are willing to do. However, you should negotiate that as *open and early as possible* with your supervisor.

4.2 Working together in a team

Here we want to discuss problems, which only occur, if several authors write different parts of the Technical Report. At the beginning of your project you can discuss the following central questions: Under which conditions are we working? What do we want to achieve? Who does what? Who has taken over responsibility for what? What are the special abilities of the group members? Who needs to learn what? Who needs to obtain what? What are our working procedures? What, how and with whom do we communicate inwards and outwards?

For teamwork the usage of the report checklist and style guide are especially important. Otherwise one group member writes "figure 3 to 8" and another one "figures 3 thru 8" and these inconsistencies do at least bother your readers.

Agree on a common structure before you start to write the text. The structure should be finer than a 10-point-structure, if possible. Changes to the structure should be communicated to the other group members as soon as possible.

Please use common fonts so that special characters and symbols will not look completely different than planned on the computer on which the final version of the report will be printed. This holds also true for PowerPoint presentations, especially for the bullet symbols.

If you are all using the same version of word processor program and the same formatting templates and the same fonts, your systems are already rather compatible. In addition, you should use the same printer driver for the (final) line and page break.

If there are computations, you should define in advance and evtl. discuss that with your supervisor, how many decimal places shall be used in the formulas.

The page numbering often does not fit together, because all group members use their own page numbering. So there are page numbers which occur more than once or gaps in the page numbering and erroneous cross-references. This needs to be adopted before the final creation of table of contents, list of figures and list of tables.

In any case, we recommend that the group member with the best computer literacy combines the text parts written by the different group members and then prints out the report, because he/she can then still correct mistakes of the others.

4.3 Advice for working in the library

In subchapter 3.5 we have already described, that working with literature is important to prove or disprove your own position, to get new ideas and to reflect the current state-of-the-art. Yet, working with literature is very time-consuming. Therefore you should try to minimize the time it needs by intelligent planning and organization. Here are some hints how to do that.

If you are not so versed with your local library, you should take part in a **guided tour**. There you learn, how the various types of literature are searched (online catalogues, microfiche catalogues, card indexes for older publications etc.), where the literature is located according to the location label and signature (on reserve shelves or in the depot, which corridor for which field of knowledge) and whether, when and how the literature can be borrowed (computer-aided loan, depot order, distance loan, short loan etc.).

Recent journals may not be borrowed in most cases. Articles from them must be copied in the library. Also, the reference copy of desired books may not be borrowed in most cases. Therefore you should start to collect **coins** in time, so that they are available when you need them in the library (e. g. in an empty **film box**). Alternatively you should buy a **copy card**. Also a box for **paper clips** is useful. You will always need them to divide several stacks of copies. In addition you should take with you at least two **well-functioning ball pens** and the **notes** collected for the next visit to the library. At home, you should preferably put these notes into a firm DIN C4 **envelope**, labeled with **"library"**.

If you have arrived at the library, you should perform all work steps in the following order:

- internal depot orders (last about 1 hour: you can take the books with you!)
- external depot orders (last about 1 day: you cannot take the books with you!)
- search for literature on reserve shelves (the books are on the shelves)
- evtl. search for literature in the dissertation archive, search for literature on microfiche, (let someone) make enlarged copies from microfiche etc.
- visit to a library presenting standards and patents to the public (the library employees will help you to find technical rules and guidelines, standards and evtl. patents)
- reading literature, making copies
- working yourself through the notes in your envelope "library" and further working on the text of your Technical Report to bridge waiting times

In the library there are normally desks available for these purposes. They can be used by anyone provided that you obey the internationally valid rule to keep as quiet as possible in a library. Please switch your mobile phone from sound to vibration alarm!

When you make **photocopies** in the library, *a very important working principle* is, that you should **note at once all bibliographical data** and the data where to find the book in the library on the copied sheets. At least you should note author and year onto the front side of the first sheet in the stack, if this is not already printed there, and write the rest of the bibliographical data evtl. onto the rear side of the first sheet of the stack. Then you should connect the stack with a paper clip, mark the author's name on the front side of the first page with a pen in a different color (because the names are printed at different positions in different publications) and then make photocopies from the next literature source or article from a journal. If you proceed like that, you will have all required information at hand, when you write the list of references and you can alphabetically sort the copies and put them into file binders quite easily.

You should not give back the literature immediately after making the copies in the library. At first you should **check** the copy stacks, **whether there are pages missing**. Only if you are sure, that all pages you wanted to copy are **completely** there in the right orientation and order and with the complete **bibliographical data** and information where to find the literature in the library, you can give back the literature with ease. Check again, whether there are still open notes in your envelope "library", which you want to deal with.

At home you can put the copied literature either onto the book pile of the appertaining chapter or subchapter or you can punch the copies and add them to your report binder at the right position, see section 4.4 "Organizing your paperwork".

With the approach described here you will save a lot of time. Besides, with this working procedure you will lay the foundation for being able to create a correct and complete list of references without having to revisit the library, just because there are some bibliographical data missing. In the time saved by this approach you can perform the other necessary worksteps more carefully and do the proof-reading more intensively.

4.4 Organizing your paperwork

In this section you will get to know a working method how to sort and store different cover sheet/title leaf versions, drafts for the structure, literature sources, brochures, encyclopediae, notes, text drafts and other materials assigned to your Technical Report. However, every project is built up a little different and therefore needs appropriately adopted techniques to perform the project work and to organize the paperwork.

Install a **report binder**, in which you will collect everything that belongs to the Technical Report. Label it at the back of the file binder with the working title of the report.

Subdivide the report binder with **dividers** made of firm cardboard, which exceed the paper sheets at the right margin and which are labeled with soft pencil. This has advantages, as long as you are not yet very deeply involved in your project. At this early time you do not yet know all details, how the information should be ordered and structured best. Therefore it may occur, that you want to label the dividers differently in the future. This can be done easily, if you use a soft pencil and a rubber.

The report binder may for example get the following **structure**:

- cover sheet/title leaf and structure versions, the latest version is at the top
- other parts of the front matter, e. g. list of figures and list of tables, evtl. with manual additions since the last printout
- printout of all chapters of the Technical Report including the appendices
- material collection (text drafts, notes and photocopies) ordered by chapter
- notes and other material which can not (yet) be assigned to a chapter
- cover sheet/title leaf versions, the latest version is at the top
- style guide
- report checklist
- to-do-list with action items (buy or do something, list of open corrections)

☞ *A copy of the current structure is also always visible lying on or hanging next to your desk.*

If you prepare a larger Technical Report and the report binder gets too full, it is better to store the preliminary printouts and the material collection in different binders.

All **documents** in the material collection, **which cannot be punched easily**, but which shall be cited (text books, reference books, figure-tables, manufacturer catalogues etc.), are stored on a separate **book pile** for each chapter, subchapter or section, depending on the amount of documents. In case of single copied pages and thin brochures, which will remain in your personal archive, you have to decide, whether these documents shall be stored on the book pile or in the report binder. The individual book piles get a **cover sheet** with document part number and title and a **divider (sheet of paper)** to distinguish literature already cited from literature which still needs to be processed. Literature which you do not need any more is at the bottom.

It happens sometimes, that you (or someone else) wrote notes **into documents** from the material collection, and you decide later, that these pages shall be included as copy into the text of the report or as brochure into the appendix. It is much better, if you do not write notes directly into documents which you have not written yourself, but if you use **note sheets or sticky-notes**.

If you have many files, you can label your printed texts and graphics with path and file name, version and date. This is also true for printed cover sheet/title leaf and structure drafts. If you want to glue figures or tables onto your printouts, an appropriate white space remains unprinted. The figures or tables are inserted into the report folder in front of the relevant page of the draft text, e. g. in a sheet protector, but not yet glued onto the text page, because until the final printout there still may be several draft versions and the gluing would be a waste of time.

The text draft should be proof-read on paper, because typing errors will then be found much better. Scripture can be read more clearly on paper than on the screen.

Moreover, text paragraphs can be re-ordered much better on paper than on screen, if there is an illogical sequence of thoughts. **Corrections** should best be marked **in red** – even in your own drafts. Corrections and editing remarks in this color are much better readable, when you enter them into the computer, as compared with corrections marked with pencil or blue or black ball pen.

Analogously to the project notebook (jotter) described in section 2.5 for the practical work steps in your project you should create a **to-do-list** for writing your Technical Report and note all items like jobs that still need to be done, literature that needs to be bought or borrowed in the library etc., as soon as the items occur. This holds also true for corrections, which you have recognized, but which you cannot or do not want to do at the moment. Prior to the final printout you should look again through the to-do-list and process all left open items.

For **design and draft design reports** it has proven to be practical, to insert the part list (bill of materials) and drawings into the appendix, if the documents are not too bulky. In the appendix/appendices the first item is the part list. The next item is the assembly drawing. The next items are the single part drawings if there are any. DIN A4 drawings can be integrated into the report without problems. Also, DIN A3 drawings can be integrated easily, if you fold them twice. Larger drawings can be copied with a zoom factor to reduce their size to DIN A4 or DIN A3 and bound into the report. In addition, you should provide these drawings to your supervisor in their full size drawn on transparency paper in a drawing roll or folder.

Folding schemes for larger drawings and images or tables, which are wider and/or higher than DIN A4 can be found in section 3.9.5 Copying, binding, or stapling the Technical Report and distribution.

4.5 Organizing your file structure and back-up copies

If you create texts for your report, the first thing that must be defined is, how you want to store your data on your hard disk. It has proven to be practical that you create a **subdirectory** somewhere in your file tree, having the **same name as the working title of your report** evtl. in abbreviated form. In this subdirectory you can store the files representing your chapters, subchapters or sections. Please select meaningful filenames, you can remember easily in the future. Independent of the operating system of your computer the following basic rule has proven its worth.

☞ *In file names for text and image files you should not use space characters and umlauts. You should only use small letters, numbers and the special characters - and _!*

This is especially important, if you are working in a heterogenous network with different operating systems (e. g. Windows, Mac OS and UNIX), since in such complex networks file names are sometimes modified when copying the files so that capital letters become small letters (first letter as capital letter, the rest are small letters), space characters are replaced by special characters (e. g. %20 under Windows) etc. If you search your files from a computer with a different operating system, the searched file evtl. cannot be found immediately. Links referring to internet or intranet pages do not work etc.

Since the **structure** is growing while you are writing the text, i. e. it is refined all the time, you should note the following information in the structure **in one line of text** at the bottom of the file

- the path and file name of the structure file and
- the exact date when the structure was last modified (the version date)

This line of text is then always printed together with the structure.

In addition, the **file names of the structure files** should get the exact date or month and year when the structure was created. To see the structure files in the correct order, you should write the year at first, then the month and then the day. Here are some examples:

```
struc-casting-system-2009-07-12.doc
struc-casting-system-2009-09-27.doc
struc-casting-system-2009-09-30.doc
```

Keep **all older versions of your structure as files on your hard disk and as paper printout,** so that you can go back to a previous structure version. This may be necessary, if you cannot get enough literature in the remaining time for the focal point of your project, which has been defined in the last structure version, or if experiments cannot be performed in time, because there are supply difficulties or defects in your equipment, or the supervisor prefers a previous version of the structure.

☞ *To see your text files in the correct order, you should specify the chapter and evtl. subchapter number in the beginning of all text file names, evtl. with with a leading zero for one-digit chapter numbers and then the title of the chapter or subchapter.*

By applying this naming convention the files are listed in the same order as in the structure and you keep a good overview. Example (please refer to the table of contents of this book):

```
01-intro.doc        (chapter 1)
02-planning.doc     (chapter 2)
31-creation.doc     (subchapters 3.1 to 3.3)
34-creation.doc     (subchapter 3.4)
35-creation.doc     (subchapters 3.5 to 3.9)
```

If you use **image files** for your Technical Report, which took some time to create, you should **not only** store these files **embedded in** Word or PowerPoint, **but also in their** original format, in the original size and original resolution and, if applicable, **in a vector format**.

You will often need the image files for your Technical Report **also in a different format** like gif or jpg or in a different size (in pixels) for publishing them in data networks, see 3.9.4. A **different resolution** may occur, if you use your images with 300 dpi for printing them in a journal, with 150 dpi for printing and binding the Technical Report and with 72 or 75 dpi (= screen resolution) in a PowerPoint presentation to get a reduced file size.

☞ *The graphic files should always be archived and passed on to co-workers and publishers in all used file formats, image sizes and resolutions. This is the best way to guarantee that you (and others) can still edit the graphic files later. In case of vector files this is even the only possible way to be able to properly edit the files later.*

Create back-up copies of all your data quite frequently! So you save the results of your own intellectual work. In addition to the files on your hard disk you should at least create two or three copies of all files belonging to your project on different storage media like USB flash drive, CD/DVD, exchangeable disk, tape, etc.

Update the back-up copies of your files quite frequently (once or twice a day). To pack your data and to compress/uncompress your files under Windows you can use e. g. WinZip. Mark the files to be stored and click them with the right mouse button, then select "Send to zip-compressed folder" in the context menu and specify a file name.

Older Windows versions did not contain WinZip. Instead, the programs pkzip and pkunzip were popular. For these pograms you could create a batch file in txt format using an editor, e. g. the file `bachelor.bat` having the following contents: `pkzip bachelor.zip *.doc *.gif *.jpg *.dxf`. Such a batch file is started by typing its file name into a shell or the execute command line in the start menu, in this example: `bachelor`.

In addition, keep the **last printout of your files in the report binder**. This is *not exaggerated carefulness*, but a working rule derived from bitter experience.

If you are **surfing much in the internet**, store your most visited pages as Favorites and delete the temporary internet files regularly, so that your computer does not collect several MB of data and becomes very slow, before you do something against it. In the Internet Explorer there are several interesting commands to do that (depending on the version).

☞ *At least after each larger internet search you should execute the following commands:*
- *Open File – Import and Export and save your Favorites in your project directory as HTML file. Save this file regularly together with the other files in your project directory.*

- *In menu Special – Internet options perform the following steps to delete temporary files:*
  a) Delete cookies, b) Delete files – once with the option Delete all offline contents, once without that option and c) Delete sequence/history.
- *From Internet Explorer 7 on just use Special – Delete browser sequence.*

4.6 Personal working methodology

In the literature there are many tips about personal working methodology. For creating Technical Reports these tips are also valid in principle. However, since we deal with "technology" there are a few special aspects. The most important rule is: Plan enough buffer times!

☞ *The effort needed for actually finishing the Technical Report is regularly underestimated (even if experienced authors make a careful, conservative estimation). Therefore estimate the required time realistically and multiply the result by two. So you will have a sufficient buffer time.*

As a student you should plan the date of submission of your Technical Report so that you go and get a task (e. g. for a design or draft design project) early during the term or semester. Then you should create a time plan (e. g. in a Gantt diagram) and regularly check whether you are still within the planned time. You should apply strict self-discipline, even if your project and the work on your Technical Report do not run smooth for a while, but you should also plan small breaks and rewards for reached milestones. Do not loose your working direction in exaggerated perfectionism. It is better to plan an editorial deadline for yourself and keep it.

You may read again the rule of thumb for time estimation for writing the Technical Report presented in section 2.1. You should begin with writing the text no later than after 1/3 of your practical project work. Then you can finish the project early enough, to have time for preparing the written examinations at the end of the term or semester.

If you want to finish the Technical Report during the semester break, you should check whether the paper shop is closed and buy paper, card board, printing ink, toner, diskettes, CDs, forms, transparency paper etc. well before the date of submission or before holidays.

Also, you should copy **files and forms** offered by your professor (like part list forms, example entries to part lists etc.) early enough, so that you do not get time problems during the Christmas break or summer break. You should apply this approach for all default documents and default procedures provided by your supervisor, i. e. also for example computations, special diagrams, work sheets etc.

If you have to search much in the internet for your project, save interesting addresses as **favorites**. You should think of managing your favorites in subfolders sorted by topic. It is useful to **export your favorites** from time to time (in Internet

Explorer: File – Import and Export). You will get an HTML file with clickable links and can enter comments and intermediate headings providing a structure to all the links. You can also use this file for **team meetings** or the next **visit to your supervisor or customer**.

During the **initial stage** you should not invest too much effort for the optical appearance of your Technical Report. Reworking the line and page break should best be done only once but thoroughly during the final stage! Formulating the text and selection or creation of figures and tables should always run parallel with writing text. Unfortunately, drawing images lasts quite long! If you postpone this to the end, time problems are inevitable!

Handwritten **notes and sketches** are absolutely sufficient in the initial stage, to overcome the first writing barriers and to get a feeling for which information should be presented in which document part. If the Technical Report is then written in a first version on the PC, you can mark text parts which are not-yet-ready. Not-yet-ready text parts may occur, if you want to re-read or verify contents later, but you want to continue with writing text at the moment. Here the *common use* of the **marking "###"** has proven to be practical. This marking can be found easily, using the function Edit – Find. Use this marking wherever you still need to look up or add details later. Also, if you are not finally happy with your own formulations and want to rework the text later, it is useful to mark the text with the marking "###". If you have found passages in your own texts, which need to be reworked, but you did not do this because of *time-pressure*, it is very probable, that your supervisor or customer complains about it. Therefore it is better, that you *do not compromise with yourself* and correct text passages you were not contempted with yourself immediately or insert the not-yet-ready marking "###" for still required corrections.

Sometime during the writing of your text you should define a **deadline**, from when on no new literature sources shall be searched and read and no new information shall be inserted. This deadline should be at about 4/5 of the time for writing the Technical Report. If you find important sources after this deadline, you can still use them, but these should be only really important articles or literature strongly recommended by the supervisor.

If you take a larger break and switch off the computer or finish working on your Technical Report for the current day, you should insert a marking before saving your currently edited file and switching off the computer. This marking should occur only once at the position in your report where you have finished your work. We use the marking "###break" for that purpose. If you open your file later, you can easily find again the position with the function "Find" of your word processor.

If you write your Technical Report in a **team**, the layout of the text, the structure and the terminology must be defined for all group members. If you do not follow this basic rule, inconsistencies may occur, which bother your readers or make it harder to read the report. Examples:

• Within one work there are different page numbering systems.

- The different document parts of the Technical Report have different document part headings (different font size, with and without underlining, with and without document part numbers) and different methods to emphasize text are applied (bold, italic, indented).
- Within one work parts have different part names in different chapters.
- The hierarchy of the document part headings is not consistent.
- The graphic design in drawings (usage of boxes, bars, lines, arrows, section lining, or area fill patterns) is not consistent.

To create a consistent report, the tools "style guide" (subchapter 2.6) and "report checklist" (section 3.9.1) have proven their value. Especially during the initial stage of a project it is often hard, to use the established technical terms in a correct and consistent way. To assure a consistent use of terminology, include all technical terms which are important for your project (evtl. with synonyms, which are also used in the literature with a short definition) into your style guide or into a terminology list. These lists or files are constantly updated with the progress of the project. If several group members are working on the report, the updates should be exchanged weekly or even more often.

The following rules are useful for *writing the text*:

- Always keep the latest version of the structure in sight (it is the "backbone" giving you guidance and support during formulating the text).
- When formulating and designing the Technical Report switch your mind to the position of the reader and look at all details from his point of view.
- Including figures and tables into your text usually improves the understandability, but it also takes a lot of time to create them. Therefore create figures and tables parallel with writing your text. If you postpone this work too much, time problems are inevitable.
- If you want to bind a brochure together with your Technical Report or cut out relevant images from it and glue them onto your copy originals, you should not write notes directly into the brochure. Better use note sheets which you lay into the brochure or Post-it sticky notes which you fix onto its pages.
- The pure typing of one page of the list of references takes two or three times as long as typing an ordinary text page. Therefore you should write the list of references parallel with the text into an own file.
- After a while you will become routine-blinded against your own formulations. Therefore give away your Technical Report for proof-reading to a person who is technically educated, but does not know the details of your special project.
- If you are not very good in spelling and punctuation, you should also give away your Technical Report to a person who is good in spelling. Evtl. buy yourself an Oxford manual of style (for British English) or a Chicago manual of style (for American English) or a similar book.
- Formulate the introduction and summary with special care. These two chapters are read thoroughly by nearly every reader!
- Plan enough time for the phases proof-reading, master copy creation and end check. This makes your report more consistent and improves its quality.

- Always note things that still need to be done on a sheet of paper, in a to-do-list or a notebook! Otherwise the risk to forget details is too high!

At the end check thoroughly, whether you have fulfilled the task and whether the report is optically "smooth". Check formal aspects with the report checklist and the style guide. Then your report will become consistent and reach a high level of quality.

After the written Technical Report is completed, you can approach the task to orally present your topic now.

5 Presenting the Technical Report

Today the best Technical Report is only useful for someone, who can present it successfully. All what matters in the professional area – in doing business or politics – is strongly influenced by **personal contact**, by the spoken word, no matter how well it is prepared in written form. Therefore, if you want to have **success**, you cannot avoid **presenting**.

5.1 Introduction

The following pages introduce to you the **characteristics, purpose, and background of presenting**, taking a lecture or technical presentation as an example. Then a systematic approach is shown that helps you to save time, money and nerves, when **planning, creating and presenting your speech**. Without that systematic approach, just with a talent to talk it is impossible to present a good lecture or technical presentation. Please keep in mind, that during a lecture (or other type of oral presentation) as well as when showing a piece of wizardry "you can only come up with things like that, which you have prepared before"!

5.1.1 Target areas university and industrial practice

You are a student just before presenting your master thesis or a postgraduate just before presenting your dissertation? You are an engineer and want to present a paper in your company or at a conference? If yes, the following chapter helps you to learn or improve presenting your topic(s) to important people like your professor, your management, or a public audience.

The following rules, hints, and tips refer to **examples from universities** like student research project reports, bachelor or master theses or dissertations, but they are also applicable e. g. for an oral status report in a meeting in a company or a presentation of a new product at a fair.

Student or engineer in practice – Do you want to slip into both roles?

Then you can probably agree with the following: What you get to know in this part of the book is **valid for everyday life in university as well as for the professional practice**. In all fields of life you are presenting your message to an audience of *human beings*, and they share the same expectations and behavior to a larger extent than usually estimated. You don't believe that? Try it out!

L. Hering, H. Hering, *How to Write Technical Reports*,
DOI 10.1007/978-3-540-69929-3_5, © Springer-Verlag Berlin Heidelberg 2010

5.1.2 What is it all about?

"My Technical Report is finally ready – presenting it won't be a problem!"

Or will it be one?

No, the lecture or presentation will not be problems, at least none which are impossible to solve!

Since you, dear readers, have created so many texts, tables, diagrams, and images throughout a long and busy period of time during the work on your Technical Report, that all this material should be sufficient for a lecture or presentation of 20, 30 or 45 minutes.

However – is that the type of information you can quite easily create a good, successful lecture or technical presentation from? **Why** do we present **lectures** at all, **when everybody who is interested can read everything** in your Technical Report? These and similar questions shall be discussed and answered in the following sections:

- Which **right to exist**, sense and purpose does **presenting** a Technical Report have in modern business life?
- **What needs to be defined in advance**, e. g.
  - which target group(s),
  - which time frame,
  - which devices?
- How do you **plan** the creation and presentation of your lecture?
- How do you **structure** your lecture, which amount of information and level of knowledge do you select?
- **What is important** to get a good grade, praise or acceptance after your presentation?

This leads us again to the question: Why do we present lectures at all? What is all the effort good for – everybody can read everything!? What are the advantages of preparing and presenting a lecture? The answer can be found in section 5.2. But at first we want to point out the advantages for you personally.

5.1.3 What is my benefit?

It is very easy – by presenting your report

- you as a person have much more influence,
- you can explain the contents of your report better and show the advantages,
- you will have more success,
- you will have more fun.

Let us begin with the fun: All presenting ladies and gentlemen will reveal (if they are honest), that it is an **exaltation of your own abilities**, zest for life, and also of having a little power, to successfully get your message across to people. Naturally speaking, this feeling does not come up when you give your first presentation. This is prevented by the stage-fright, all presenters have at first, and which does never disappear completely.

But with growing routine you will get a feeling of **contentedness and satisfaction**, to "communicate" with an audience ("to exchange something with the audience and then to have it in common"). You will have fun to get a welcome change from your at times boring workaday routine, which helps you develop yourself in terms of communication and knowledge, i. e. which provides you with personal and **professional success**.

With your presentation the **contents becomes more vivid**, because **due to your personal appearance** and your freely spoken explanations it gets a much more intensive impact. To express it in easier words, your report mostly addresses "the head", the brain, the intellect of your readers, while the presentation reaches "the gut", the emotions, the unconscious in your audience. Only if there is a complete and complementary impression of emotion (first) and intellect (second), you will provide a holistic and impressive experience of your work and your person standing behind all that work. There is an insight in psychology: When the feeling says No, there are no facts and arguments which are good enough to convince your audience. You have already worked on your report for quite a while, all this effort shall result in a good grade, more money in your purse, and better career chances! Wouldn't that be a success?

5.1.4 How do I proceed?

The basic attitude you should take when starting to work on your presentation is:

"What does my audience need and wish to get?"

The **presentation** of your report is **something completely else than** just summarizing or even reading parts of your **written Technical Report**! It is clear that all the effort and highly qualified work of many days, weeks, months, or in case of dissertations of years cannot be condensed to the ridiculously short time of 20, 30, or 60 minutes. Therefore you can only select a small fraction of your Technical Report, sort of a peek through the keyhole. But which peek do you select? For this decision you need three important phases:

Phase 1:   **In your mind**, try to get as much **distance from your report** as possible.

Phase 2:   Slip into the position of your audience including professor(s), your boss and other experts and try to **anticipate** their existing **knowledge, attitude** and **expectations**.

Phase 3:   You **try to meet these expectations** of your audience as well and
complete as possible.

You will ask yourself: "Is all that really necessary?", "How can I reach that and
how large is my effort to succeed?"

You will find the answers to these questions on the following pages, together
with basic rules, hints, tips and tricks, which will help you to avoid mistakes, save
time and be successful. Become surprised, step in!

5.2 Why presentations?

In this section the **properties of a lecture or a presentation** shall be pointed out
and the differences between lectures or presentations and Technical Reports. In
addition, **presentation targets** and **presentation types** are described. But at first
we want to introduce a few definitions of terms which are often mismatched.

5.2.1 Definitions

to present	introduce, show, make clear and present
	(present as a gift, present as present tense)
lecture	longer oral speech (20 to 60 min)
	about a topic or on a certain occasion
	with **transferring knowledge**
	and/or **influencing opinions**
presentation	lecture of medium length (10 to 15 min)
	about a project or a product
	with **transferring knowledge**
	and **influencing opinions.**
statement	short lecture (5 to 10 min) on a topic,
	for example **expertise,**
	point of view,
	own opinion.
information	1. shortest lecture just for informing other persons
	2. other word for **transferring knowledge**
	(should always be objective and neutral!)

Naturally speaking these **terms cannot be divided from each other very
sharply**, but they have **common properties**. The **lecture** is the **master type** of all
speech forms and it is most elaborate to prepare and to present. All other speech

forms can be derived from the lecture and have more or less elements in common. Therefore we should get to know the properties of a lecture especially well, even if it is abbreviated in a presentation or a statement. *So, if we speak of a lecture, we also want to address the derived speech forms as a whole or in parts.*

5.2.2 Presentation types and presentation targets

Before we speak about presentation types and presentation targets we want to explain the term "to present" in three aspects: *First*, to get across the **contents** of a lecture, i. e. to make it **present** for the audience, a good speaker should *second* **be present** as a person, e. g. via charisma and engagement, and *third* he should give his **lecture** to his audience **as a present**. Presents are exchanged between friends. Therefore, at least during the lecture and the following discussion, there should be a **basic attitude of friendship between speaker and audience** (a friendly relationship aspect). We want to define four important presentation types, **Table 5-1**.

Table 5-1   Presentation types

Presentation type	Properties
1. lecture	• The preference is pure information! • Objectivity is a must. • Main contents: technology, presented appealingly. • The audience is addressed so that they understand the contents easily, i. e. indirectly.
2. persuasive presentation	• The preference is persuasion with technical and non-technical arguments. • Technical objectivity only as far as it is needed. • The presentation must be especially appealing, persuasive, and eventually amusing. • The audience is addressed very specifically and directly.
3. technical presentation	• The preference in most cases is influencing the audience. • But still a lot of technical information is provided.
4. occasional speech	• The preference is influencing and amusing the audience and appealing to their emotions. • Only a little technical information is provided (minor point).

The transitions between these presentation types are fluent!

Presentation targets
A lecture, presentation or speech always has three targets, each with a different weight:

Target 1: Informing
Technical knowledge   • *shall be documented*
                               (saved in texts, figures and images) and
                             • *transferred*
                               (communicated to other people)

Target 2: Persuading
Persuading the audience
                             • of the quality of the provided knowledge
                             • the efficiency of the delivered work
                             • the competence of the candidates and/or
                               co-workers involved

Target 3: Influencing
Influencing the audience to act:
                             • granting money
                             • buying
                             • continuing the project
                             • positive decision
                             • good grade

5.2.3 "Risks and side effects" of presentations and lectures

In all technical professions there are still written and oral forms of instruction and communication, where "written" includes the text-based communication via data networks. Why is none of these communication forms sufficient? The answer is: Both communication forms have advantages and disadvantages, comparable with the "risks and side effects" which you have often heard about. Let us look at both communication pathways with all their strengths and weaknesses.

The written communication (documentation) comes up as notes, reports, manuals, electronic messages or written offers.

The oral communication contains arguing and presenting in a talk, a meeting, or in a presentation or lecture.

Table 5-2 shows that both forms have about the same number of strengths and weaknesses. Therefore you need both forms in your professional life, and therefore this book does not only show you how to build up a Technical Report but also how to present it.

Table 5-2   Advantages and disadvantages of written and oral communication

	Advantages	Disadvantages
Written communication: documenting, i. e. from "note" to "offer", incl. e-mail	Independence of time, editing is possible at day and night: all contents is visible; „reader" can control his/her reading and read again: contents is clearly set; „writer" can work with concentration, paragraph by paragraph, take breaks; quality can be controlled well. Verification is always possible.	Reader cannot ask back, impersonal, inflexible; no control of the effects and which reader reads under which circumstances, no feedback; What is written can hardly be taken back/smoothed/emphasized.
Oral communication: arguing, presenting, i. e. from "talk" to "lecture"	A good speaker is enjoyable for the audience; speaker has contact, gets feedback, can make convincing use of his personality; can correct himself, can react flexible, can make use of/control moods and emotions (up to demagogy and manipulation); spoken word, visible images and physical presence of the speaker altogether result in a strong impression.	Missing reproducibility, need for physical/psychological strength, strong nerves, active knowledge, discipline, intellect, quick-wittedness, knowledge of people, sensibility, eloquence, good manners, positive impression, charisma, good form on that day! Disturbances because of comments or questions or intentional interferences; possibility of misunderstandings, risk of errors in reasoning is always there, constant time-pressure, stage-fright, self-doubt.

The question asked at the beginning "Why presentations?" can be answered now:

☞ *To **convince** people of technical contents and to **influence them in the desired way** you need more than written communication. Human beings are no "scanners" (which would be sufficient for the pure perception of the information contained in your report), but as decision-makers they are also characters with head and gut. In addition to reading your report with the "scanner function" the senses "**Hearing**" and "**Seeing**" want to be addressed. Last but not least the **intuitive feeling of your personality** via voice and visual impression is also essential to convince and influence your audience as far as possible.*

Voice and images are the **basic elements and strengths of a lecture**; they make the oral presentation inevitable, if there is something important to decide. That is why speeches in parliament, oral lessons at school and in vocational education as well as lectures in study courses play a dominant role. And finally another, not unimportant aspect: The personality of the speaker is not only transported via the voice, but also via nonverbal signals like **erectness, facial expression, gestures and charisma.** They have a **strong influence on the audience**, which is missing to any report or book. Therefore, when standing behind the speaker's desk we should know these relations, processes and results and build up our lecture completely different than our Technical Report.

☞ *The **lecture or presentation** is a **new creation** – based on the report, but **with completely different focal points, images and elements of style** – if it shall be successful.*

Now you have enough background knowledge and motivation to plan your presentation.

5.3 Planning the presentation

Yes, a creative chaos has its sympathetic sides, but without planning you will soon run into time-pressure with respect to the date of your presentation. Therefore the following pages describe the work steps to prepare and present a lecture and their time consumption.

5.3.1 Required work steps and their time consumption

Figure 5-1 shows a network plan with the required work steps ordered by their time-sequence. This plan recommends eight steps, which need quite a different amount of time. And these tasks do not run strictly one after the other. Please refer to the Gantt diagram, **Figure 5-2**. All tasks and recommendations for your time-planning refer to the essential presentation, which needs to be as perfect and successful as possible, like presenting your master thesis or applying for a job or project continuation.

Naturally speaking, "normal" lectures created in day-to-day business can be created with more freedom and less time consumption as your routine grows. The proposed **time-frame** assumes, that you have available **two workweeks with two weekends**. That means the preparation of your lecture takes place part-time during the week and in the spare time during the weekends.

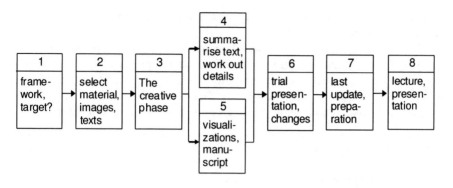

Figure 5-1 Network plan to create and present a lecture

Worksteps	Time consumption	1st week	2nd week	1-2 days
1	define presentation framework and target	▪		
2	collect/select or create material, images and texts	▬		
3	The creative phase: gain distance! develop a structure		▬	
4	summarize the text, work out the details		▬	
5	create (more) visualizations, write the manuscript		▬	
6	trial presentation, include changes			▬
7	last update of the lecture, preparations in the room			▪
8	hold lecture or presentation			▬

Figure 5-2 Gantt diagram to create and present a lecture

Figures 5-1 and 5-2 as well as **Checklist 5-1** need some explanations:

The time consumption is specified in "gross days". According to the authors' experience this situation is the most common, i. e. **working out a lecture** during day-to-day business **always requires working time and spare time**. The spare time must be detracted from private life, but a successful presentation can be the base for a jump in your career.

The **weekends** are mainly needed **for step 3** "The creative phase: gain distance! develop a structure" **and for step 6** "trial presentation, changes". Especially for **step 3** you will need some calm and lonesomeness to unfold your creativity, which is not provided at most workplaces. The Gantt diagram shows recommendations for the required time you should take into consideration, **Checklist 5-1**.

Checklist 5-1   Time consumption to create and present a lecture

Step 1:	define presentation framework and target	1 day
Step 2:	collect/select or create material, images and texts	3-5 days
Step 3:	the creative phase	2-3 days
Step 4:	summarize the text, work out the details	2-3 days
Step 5:	create (more) visualizations, write the manuscript	3-4 days
Step 6:	trial presentation, include changes	1-2 days
Step 7:	last update of the lecture, preparations in the room	1-2 hours
Step 8:	hold lecture or presentation	20-60 minutes

The tasks described above (Steps 1 to 8) are explained in detail in the next sections.

5.3.2 Step 1: Defining the presentation framework and target

Planning your presentation consists of step 1 "define presentation framework and target" and step 2 "material collection". With step 1 you build a solid foundation for your presentation by defining all important **presentation conditions, the target group and the information targets of your lecture** before executing the next worksteps. This helps to prevent a failure and to save time. You should ask for the following presentation conditions and targets, discuss them in your team or define them for yourself, **Checklist 5-2**.

Checklist 5-2   Presentation framework (presentation conditions, target group, information target)

- What do I talk about? (topic)
- What shall be the title of my lecture?
- What kind of lecture will I present? (presentation type)
- Who will listen to me? (audience, target group)
- What is the occasion to present my lecture?
- What do I want to achieve with my lecture? (presentation targets)
- Why do I want to achieve that? (purpose)
- Where do I present my lecture? (surroundings, room)

Clarify these questions in detail (if possible in written form). Then you will feel a little safer, and the lecture cannot be a complete failure any more. In the following we want to look at these questions a little closer.

What do I talk about? (topic)
Very easy – about your Technical Report!

However, the title of your lecture may differ from the title of your Technical Report, e. g. if the latter is too long (occurs often) or if it sounds too complicated for the title of a lecture or if a better title came to your mind.

In professional life you will hold lectures about various, important topics. Then the topic and contents should be agreed upon as exactly as possible with your customer, or boss or chairperson of a conference or organizing staff of a fair. This includes a (written) **definition of**

- title, subtitle
- audience (knowledge level, interests, expectations)
- exact timeframe without/with discussion
- topic and presenter of the previous and next lecture
- presentation type, room size, equipment (desks/chairs/boards), devices (projectors/flipchart).

For your career it is very **important to arrange things with**

- your **boss**
- your **supervisor, professor or customer or the organizing staff of a congress, conference or fair**
- your own **sales and marketing team leader**!

Why should you speak with the sales team leader and the marketing team in any case?

An example:
Your company a well-known car manufacturer, has developed a super-fast sport coupé and wants to introduce this marvel with an expensive PR show at the next fair. 14 days before the fair you (being a young, enthusiastic development engineer) present the most important motor and chassis details on a conference. Among the experts in the audience is a representative of the press who launches your insider knowledge big on the next day. Which consequences would that have for your company, for you and your job?

But now let us go back to the presentation framework, that needs to be clarified.

What kind of lecture will I present? (presentation type)
Is a pure **technical presentation or** a **special persuasive presentation** more appropriate? Does the audience expect pure technical facts of your report or do they want to get insights they did not have before? Do they want an **overview or details**?

How do you mix facts, influencing and emotion best, maybe even with a pinch of humor? This depends on your own direction and information target, but also on the audience and the occasion to present your lecture. Therefore you should also clarify the following four questions.

Who will listen to me? (audience, target group)

Are they your professors or supervisors? Are they engineers, bankers or journalists, are they experts or unknown visitors, are they your boss and other executives or are they colleagues from sister departments? You should best prepare your lecture depending on these people or – if you do not have better prior information – for a colorful mixture of people. How you can do that will be explained together with the keywords "amount of information" and "knowledge level" (see **Tables 5-4 and 5-5** as well as **Figure 5-5**).

What is the occasion to present my lecture?

Will you present your master thesis or your dissertation? Is it a pure report for colleagues or a presentation for customers? Do you want to persuade investors (banks, scientific funding boards, federal ministries) or a working committee for standardization or judges in a trial? You will have to adopt your lecture to these occasions more or less, if it shall be successful – and that is why you present it!

What do I want to achieve with my lecture? (presentation targets)

This question deals with your presentation target(s). This means, whether and in which mixture you want to achieve the following effects:

- **Transfer of technical and scientific knowledge** (to make your audience cleverer) or
- **creating a positive impression** (of your technological or scientific approach, of you or your company) or
- **creating emotions in your audience** (in a desired direction).

Naturally speaking, you do not follow these information targets too obviously, but smart and carefully – Your audience should not realize your tactic, at least not consciously. As long as your approach does not exceed to crude manipulation, no one will complain about a **decent rhetorical tactic**.

Why do I want to achieve that? (purpose)

Is everything allowed that advances the purpose???
Let us be careful and say: Sometimes yes!

An example:
The future funding of your whole project group depends on your lecture. It is a "To be or not to be"-affair for the jobs of your colleagues and maybe for your job as well – isn't that worth to invest a lot of effort?
Such purposes in mind you will probably be willing to pull out all legal stops to be successful, won't you?

This includes to decide, how much you want to convince, influence or amuse. These elements of a good lecture must be mixed well and used with care. That requires a good preparation, because mixing these targets ad-hoc does only work for a few routiniers. Preparation means to think about, define and test things.

Where do I present my lecture? (surroundings, room)

Print out a description how to reach the presentation room and clarify room capacity, desk layout, needed media (board, flipchart, overhead projector, computer and beamer, moderation kit, felt pens) as well as evtl. catering for the audience and book the room, if necessary.

To explain the step 1 "define presentation framework and target" and the consecutive steps 2 to 8 more concrete, an example shall accompany us from now on, as far as it is possible to describe it within the framework of this book.

Note: The student's research project, the following example presentation is derived from, is listed in the list of references. The conference and the lecture did never take part. The "Institute for Welding Technology" does not exist at Hannover University. The workflow is just invented to describe the basic approach to you. The example presentation will be in italic letters from now on.

The **title of the Technical Report** in our example is:

Improving health and safety at work when welding
by using burner-integrated suction nozzles –
Effectivity and quality assurance
Student research project at Hannover University

The Technical Report that describes the student research project consists of **135 pages** and **shall be presented in 20 minutes**. How can this be done?

The author courageously starts and clarifies in **step 1**:

- **Topic:**   *The student research project named above shall be presented as a paper on a conference about welding technology.*

- **Title:**   *"Burner-integrated suction when welding"*
                    This title is clearer, shorter and better memorable than the title of the report, thus more attractive and still correct.

- **Audience:**   *Head of the institute (professor), supervisor, fellow-students, a person working in industry, a journalist, other participants with unknown technical background.*
                    This audience has various knowledge levels, interests and expectations, which must be taken into consideration skillfully. Nobody should be overstrained or bored!

- **Timeframe:**   *20 minutes lecture with additional discussion.*

- **Previous lecture:**   *"Health and safety in crafts"*
                          (Person from the employers' liability insurance associa-
                          tion)
- **Next lecture:**   *"Automated flame cutting"*
                          (Person from the company Messer-Griesheim)
                          To avoid interferences and contradictions, it is necessary
                          to contact these presenters and to harmonize the contents
                          of the three lectures!
- **Presentation type:** *Technical presentation*
- **Room:**   *Lecture room 32, 30 seats, desks*
- **Media:**   *Overhead projector and beamer available*
- **Agreement**   *Planning discussion: Rough selection of the contents,*
  with professor   *strategy to meet the persons from industry and the.*
  and supervisor: *journalists*

Building up on this presentation framework the speaker designs her plan of action. She especially thinks about the **mixture of technical information, convincing and influencing** taking into consideration **the following criteria**:

- **Target group:**   *"The audience described above is very heterogeneous.*
                      This requires to create a balance between experts (pro-
                      fessor, supervisor) and the other listeners."
                      This balance can be created with a smart selection of the
                      contents (see 5.3.4 "Step 3: The creative phase").
- **Occasion:**   *Seminar or conference "Current state of welding techno-*
                      logy", Obligatory for students, public event. Back-
                      grounds are also to gain funding from industry for the in-
                      stitute and to improve the image of the institute.
- **Targets:**   *The lecture shall mainly serve to instruct the audience*
                      about the findings of the project (70 %), but it shall also
                      show the competence of the institute (20 %) and introdu-
                      ce ease/humor into the uninspiring contents (10 %).
- **Purpose:**   *Main purpose: Good grade!*
                      Desired side effects:
                      – Awake interest in industry and crafts.
                      – Get a good echo in scientific journals.
                      – Bring knowledge and motivation to the students.

Step 1 is now completed. Let us continue with step 2 "Material collection".

5.3.3 Step 2: Material collection

To work on **step 2** two situations must be distinguished:

Case 1: The **material** (the contents) of the planned presentation **already exists**, e. g. in the Technical Report.

Case 2: The **material must** be **collected, found, created**.

In **Case 1** we have the **complete report available.** Then it is the task to **select the information suited for the presentation** from the rest of the information contained in the report. Suited means that the information is **essential, meaningful and representative**. Thus, in case 1 the material collection mainly consists of the selection of contents. Please refer to section 5.3.4 "Step 3: The creative phase" for more details.

In **Case 2** we do not have a ready report. This is more difficult. The **material must be collected or found**, e. g. in textbooks and binders in your office, in archives, libraries, or the internet. This may take about **3-5 days**. If this time is not sufficient, you are not enough an expert and cannot present this topic convincingly. Someone else should hold the lecture.

If you have found and printed or copied enough **material**, you should **sort out** the less important material (texts and images) and put it aside as a reserve.

The **other material** should be **sorted by**:

- literature sources (consecutively numbered),
- topics (see section 2.4.4, Checklist 2-6, 3rd step)
- and/or author(s).

To do this systematically you should create **a file/an index.**
This completes the routine work for your presentation.

5.3.4 Step 3: The creative phase

In **step 3** the most important, **most difficult, and most exciting phase** on your way to a successful presentation takes place:

- **Gaining distance** to the contents, gaining an overview
  from the perspective of the target group (audience) and
- finding (creating) a **structure**, which represents the relevant contents and
  meets the expectations of the audience.

That is no easy trick. In spite of the fact, that you are an expert in your topic and that you have worked out and documented it, in your mind you step next to yourself and imagine the situation of your listeners. Meeting their expectations you develop a **concept for your presentation**, a **basic outline of what you want**

to describe, maybe with a special gag and with a **dramaturgy** – as much science as needed, as much clearness as possible ...

How far you (may) deviate from your Technical Report and its structure only depends on the targets and purpose of your presentation or lecture. This may result in a completely new structure of the information, which strongly influences the lecture and makes it successful.

This creative phase may last **2 to 3 days** – it makes your lecture special. Let us look again at the speaker in the example "Burner-integrated suction when welding".

*The **Technical Report** on the student research project has the following **structure**:*

1. *Introduction*   *(2 pages)*
2. *Fundamentals (Welding methods, pollutants and their removal)*   *(29 pages)*
3. *Optimization of the burner-integrated suction*   *(24 pages)*
4. *Welding tests with the optimized suction*   *(68 pages)*
5. *Summary and outlook*   *(4 pages)*

This structure of a Technical Report is sound and informative, but not well suited as a structure of a lecture or presentation. Why?

A **structure of a presentation** needs **a few, short**, inspiring **items**, which can be memorized well and have a dramaturgy.

*Our speaker creates the following **structure of the presentation**:*
Introduction: *Health vs. seam quality? ("wow effect")*
Main part: *– Suction techniques (current state)*
                    – Burner-integrated suction
                    – Design optimization
                    – Tests to measure the weld smoke (measuring the isotachs)
                    – Tests of the seam quality (destructive and non-destructive)
Conclusion: *Summary, answer to the "wow effect"*

A Technical Report or another technical topic can be transported like this or in a similar way in 20 minutes. Naturally speaking, such a concept should not result in a sales presentation. It depends on your instinctive feeling here not only to convince the professor who will give you a grade by an **obviously serious scientific approach**, but also to raise the knowledge and create a positive attitude towards your topic in all the other people in the audience by **an understandable and clear approach and your personal charisma.**

Realizing the structure

Now it is necessary to realize the structure, i. e. to make it vivid with **statements and information, texts and images.** Everybody in your audience from the professor or experts to the accidental guest shall "take something home" from your presentation – all of them shall think that your presentation was a present, at least partly.

Can this task be fulfilled? Yes, it can – <u>partly</u>! Partly means: **Everybody in your audience** will get **his part** of the contents of your presentation, but not all the time. The secret is to smartly distribute the contents of your lecture depending on the individual expectations of the different persons or groups in your audience:

- Widely known facts for all listeners ("Well-known")
- Complex scientific information for the experts ("Mazy")
- Understandable new information for the majority ("New")

The third-rule

As a presenter you will seldom have an ideally homogenous audience, because even among experts there are big differences in their knowledge the more special our modern knowledge becomes. And you want to prepare your presentation only once and re-use it for more than one occasion, if possible. Therefore you should apply the third-rule, **Table 5-3**.

Table 5-3   Third-rule (trisection) to design a presentation

Part	Explanation	Purpose
1/3 "**Well-known**"	Directly affecting, daily things, everybody knows.	Common base, self-affirmation to the audience
1/3 "**Mazy**"	Special facts and details, which are only understandable for the experts.	Raise of knowledge for the experts (insiders), image improvement for the speaker
1/3 "**New**"	Scientific and/or technical information on an appropriate knowledge level, understandable for the majority	Raise of knowledge for the majority, evtl. even for people who are not acquainted with the subject area.

The **third-rule** may look a little tricky, but it is the **best way** to find the clue a **very inhomogeneous audience** needs. It has **often proven to be practical** – find your own judgment, try it out! To realize the third-rule you cannot just present the parts one after the other according to the fractions, it is a little more differentiated. A presentation should not be split into three exact thirds, but it has three parts. **Realizing and using the trisection**: This trisection can be found in the biology of daily life and in technology, as the following examples show, **Figures 5-3 and 5-4**. This trisection is realized in a technical presentation by smartly varying the knowledge level, to meet the individual expectations of as many people in the audience as possible, **Table 5-4**.

Admitted – the exact distribution of the main part seems to be a little artificial and farfetched, but it makes sense as an orientation and recommendation, especially due to the section into thirds or sixths resp. according to the third-rule. What is the secret?

A. Biological processes

The biology of human beings needs a certain optimum sequence for all important processes in the soul, intellect and body.

Example 1:   **Eating**

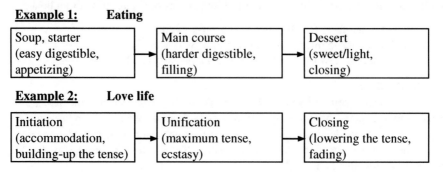

Example 2:   **Love life**

Figure 5-3 Processes compared with a presentation (biology)

B. Technological processes

Example 3:   **Combustion engine**

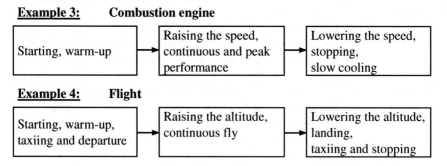

Example 4:   **Flight**

Figure 5-4 Processes compared with a presentation (technology)

    After contact preparations and contacting the audience, which are described in section 5.5.1, the presentation should start with the phases **"presentation target and course of action"** and **"introductory examples"** to give an overview and build-up an information base for **everybody in the audience**. This shall activate ("warm-up") the audience and raise their interest for the topic of the presentation.

    Then **scientific and/or technological information** on an appropriate knowledge level should be delivered, **which is understandable for the majority of the audience** – the first sixth of the **"New"**. Before the experts in the audience become bored, you **refine the information** and present the **"Mazy"**, i. e. more complex connections, details and specialties, which can be understood only by one or a few **experts** e. g. your professor and his colleagues. However, in this part of the presentation even these experts should be able to still learn something.

Table 5-4   Design of a technical presentation

Phase	Time	Contents	Information target and fraction according to third-rule
Introduction	about 20%	contact preparations and contacting the audience, presentation target and course of action	Warm-up
Main part	about 70%	introductory examples, overview, information base	1/3 Well-known
		raise of knowledge	1/6 New
		refinement, details, specialties	1/3 Mazy
		aggregation, conclusions and evaluation	1/6 New
Conclusion	about 10%	summary (outlook)	Closing

During the phase "Mazy" the **non-experts**, i. e. the majority of the audience have growing difficulties to follow the presentation. They stay in respectful amazement of the knowledge and brilliance of the speaker or **partly turn off their conscience**. Before the high knowledge level starts to completely frustrate your audience, you go **back to an appropriate knowledge level** and present the last sixth **"aggregation, conclusions and evaluation"**. Now the majority can follow your presentation again. Here the non-experts also raise their knowledge and insight, so that the presentation was worth listening for them, too.

A **short**, but meaningful **summary** of the most important findings and key aspects of the contents of the presentation, evtl. a short outlook, but **no new**, essential **information** and friendly sentences form the **Conclusion** of the presentation.

This **juggling** with three balls with respect to the knowledge level is the secret of a good speaker, if he/she wants to **satisfy all members of his audience**. Everybody in the audience shall have learned something, what he/she can write into his/her diary. No one should think or write: "I knew all that before" or "All that stuff was too complicated" – This would result in: "Much ado about nothing!".

In a graphic model this just described design of a technical presentation could look like it is shown in **Figure 5-5**. In this dramaturgy everybody in the audience has a benefit from listening to your presentation. After a quick, steady increase of the knowledge level the majority learns something new, and then the experts are informed on a high or very high knowledge level. In this phase the knowledge level peaks return frequently to the level of the majority, to keep up the conscience of the non-experts and guarantee transparency. Shortly before the end of the main part the knowledge level is quite high.

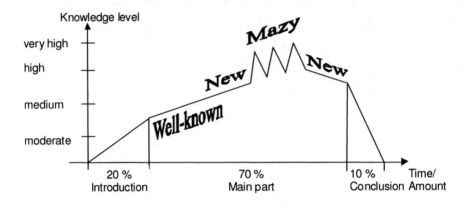

Figure 5-5 Dramaturgy model of the knowledge level

Everybody shall take something valuable home. That is also important for the usually following discussion, which is useless for the non-experts without a certain level of understanding.

The speaker in our example "Burner-integrated suction when welding" performs the trisection for her presentation, **Table 5-5**.

In subchapter 5.4 "Creating the presentation" you will find more details how this theoretical model can be realized in practice.

Table 5-5   Trisection and assignment of contents in the example presentation

Main part (with time planning)	Phases (with time planning)	Contents
4 min. Introduction	Contacting the audience, presentation target, overview	• Is the seam quality more important than health and safety at work? (provocative question, raises tense) • Examples of the problem (clearly described)

14 min. Main part	4 min. "Well-known" (state-of-the-art)	• Removal of pollutants during inert gas welding
		  – Example suction in a working hall
		  – Example suction at a workplace
		  – Example suction with a hood near the burner

	2 min. "New" (improvements)	• Burner-integrated suction (process, parts, problems) • Optimization of the VACUMIG welding pistol (suction geometry, seam quality)
	5 min. "Mazy" (special knowledge)	• Suction tests (test details, combination of suction rings, test results) • Welding tests (test details, program plan, example of measuring protocol, test results)
	3 min. "New" (consequences)	• Evaluation of suction tests • Evaluation of welding tests • Error calculation • Technical consequences
2 min. Conclusion	Summary (outlook)	• Summary of results and findings • Future work

Especially the **inevitable trial presentation** will relentlessly show, what can be done and what cannot be done in 20 minutes.

5.4 Creating the presentation

The concept is ready – now it must be elaborated! The structure with its skeleton for the contents and the time consumption must be transformed to a clear, successful presentation.

5.4.1 General recommendations for designing presentation slides

In this section you will find general hints for designing presentation slides and – scattered across the text – tips for the practical making of slides with PowerPoint 97 or 2000 or 2002, followed by tips for Open Office 2.0 Impress. This book can only show the general direction. You can find more details in the online help, the literature in the list of references and other respective literature.

Basic layout of the slides
At first you should select the **basic settings for the slide and title master**. In PowerPoint you can do this in the menu **View – Master – Slide master**. Here you can influence the layout of the list or text object resp., which is contained in the preset slide layouts List, Two-column text, Text, Text and ClipArt etc. On the slide master you can select

- font type, color and size (Format – Character),
- language (Special – Language),
- tabs,
- form and color of the bullets in lists (Format – Lists) and
- line spacing (Format – Line spacing).

In any case you should move the **left margin of the text area and the title area**, to gain **space for your structure**. Now you can create **structuring lines or other graphic objects with the tools of the Drawing toolbar**. To modify the background, you should go to the menu Format – Background. There you can select background colour, images, area fill and the logo of your company, institute or department.

In PowerPoint these settings are separately defined for the title slide in menu View – Master – Title master, i. e. you have to repeat the settings there.

If you recognize later, when you create your presentation slides, that something basic is (still) wrong with the text fields or other systematically repeated elements, you should preferably change that on the master slide.

Font type and size
It has proven to be practical to use fonts without serifs like Arial. The font size should not be smaller than 18 points, the larger the better.

Use of color
Colors can occur as text color, background color and object color in images. If used with care, **colors help to specify, emphasize, structure, order and differentiate**. Colors control reactions, stimulate imagination and initiate **emotions**, memories and associations. Use the same colors for the same things and use them **consistently**! If colors are used sparsely, they serve as "attraction" and help to emphasize important details.

However, color **reduces the contrast** between background and foreground and the recognizability. You can recognize best black on white. Text written in dark blue, dark green and dark red on white must already have a 1.5 times larger font size and lines must be 1.5 times thicker to achieve the same recognizability. Too many colors without explanation foster the stimulus satiation and burden the overall impression.

Readability check

To check the readability you can show a critical slide in screen presentation mode. Take a folding rule and measure the width of your screen, multiply this value by 6 or according to DIN 19045 even by 8, and then stand in front of your screen at that distance. Alternatively print out this critical slide on DIN A4 paper, lay the printout onto the floor and look at it while standing upright. If you can read everything, it is OK.

Presentation framework slides

It has proven to be practical, that every presentation has three framework slides to assure transparency for your listeners:

- The **title slide** (presentation title, name of the speaker, name of the institute, company or department),
- a **structure slide**, showing the structure with normally formulated, long items (can be replaced by a structure written onto a flipchart sheet or a poster at the wall) and
- the **termination slide** (contains a graphic or a photo to encourage the discussion and shows the contact data of the speaker).

Transparency assurance

Transparency assurance is one way to consecutively display the **current items in the structure** like introduction, main part with subordinated items and summary/conclusion on the slides, **Figure 5-6**, and thus assuring the transparency of the presentation.

A modern, self-confident audience always wants to know, where they are in the flow of the presentation. Then they can better prepare for the current aspect of the topic and do not start floundering through your presentation.

The structure of the presentation should be **always visible** to your audience to provide orientation. The speaker should clearly point out, where he/she is within this structure from time to time. This can be achieved by

- repeated reference to the structure and depicting the current position by ticking off the finished items at the flipchart, on a poster or a (white-/black-)board or
- by a structure in a sidebar (= area on the left or at the top, max. ¼ of the total slide area, separated from the visualization, showing the preferably very compact structure, having selected the current item). This is the only option, if in the lecture room there are no boards, no flipchart and no possibility to fix a poster to a wall.
- The structure can also be handed out to the audience prior to the presentation.

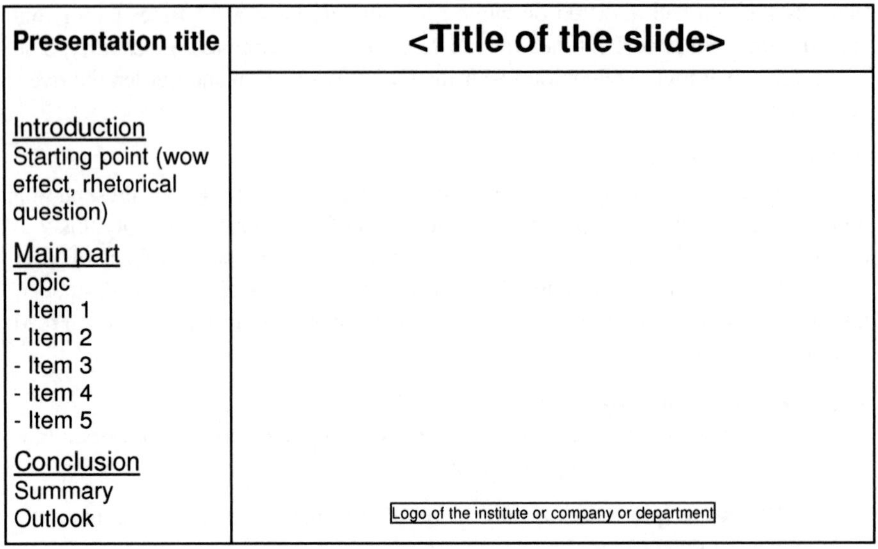

Presentation title	<Title of the slide>
Introduction Starting point (wow effect, rhetorical question) Main part Topic - Item 1 - Item 2 - Item 3 - Item 4 - Item 5 Conclusion Summary Outlook	Logo of the institute or company or department

Figure 5-6 Transparency assurance by a structure in a sidebar (schematic)

The prerequisite for this method of transparency assurance by a structure in a sidebar is a very compact structure with short keywords!

Slide design, entering text, including images

For slide design keep in mind the general principle **"The fewer, the better!"**. Avoid everything on your slides and in your beamer presentations, which is dispensable and can lower the level of conscience of your audience, e. g.:

- constant repetition of the presentation title,
- constant repetition of your name and the name of the institution,
- constant repetition of the date and occasion,
- constant repetition of copyright notes,
- constant repetition of your personal slide management notes (e. g. "Slide 8 of 31") etc.

Does the „constant repetition" bother you as well?

Admitted, this can easily be created with the computer, but the audience – as curious as it is – will (have to) read all these things again and again, just to check whether or not something has changed ...

To enter text in PowerPoint you should use the **predefined text object area** with the bullets. In Word you can enter a dash with Ctrl + minus sign in the number keypad. In PowerPoint you can use the key combination Alt + 0150. To change the indentation of a list you can use the buttons ⇨ and ⇦ in the toolbar Format. Now you can **send your slide texts to Word** with the function File – Send to – Microsoft Word and create a handout or script from the slide texts. Another useful option, which only works, if you use the predefined object areas, is the **fast entry of text slides in the structure view**. By typing Enter you create a new slide. If you enter text now, this will create the slide title. Press Enter again. A second new slide appears. Select this slide and indent it to the right with the right arrow key. Now you can enter text for your first slide. Subordinated items are indented with the right arrow key again.

There are many possible methods to **insert images**, which can only be described rudimentally here. For example, right click a slide and in the context menu select Slide layouts. Select the layout Text and ClipArt, enter your text on the left and insert a ClipArt graphic on the right.

To insert a **screenshot**, create a new slide as described above, select the object area for the ClipArt and delete it with the Del key. Change to the other program, "snapshot" the scene with the key combination Alt-PrtScr, change to PowerPoint and insert the screenshot with Ctrl-V. Click on the inserted image and scale it as desired. Grab the image only at the corners to scale proportionally. You have numerical control, if you right click the graphic, select the menu item Format graphic and change to the Size tab.

If you want to cut an image, you can use the cutting tool in the toolbar Graphics. This makes parts of the image invisible, but they still exist. You can **change the image size** with menu items Format – Graphics, Size tab.

Entering the structure at the left margin of the slides

The structure at the left margin of the slide can best be created as a text field (toolbar Drawing). Insert the field, enter your structure, format the text as desired and move the object to the desired position. If you click into the text field, it is selected and surrounded by a thick section-lined border. If you click on the border, the section lining disappears and the border becomes dotted. Now you can position your text field with the arrow keys. An even more exact positioning is possible, if you press the Ctrl key, keep it pressed and then use the arrow keys. If the text field border is dotted, you can copy a text field on one slide with Ctrl-C and insert it into the next slide at the same position with Ctrl-V. Also you can delete it with the Del key.

Footers

You can often see, that presentations have footers containing the presentation title, the speaker's name and a page numbering, like "8" or in extreme cases "Slide 8 of 23". Your audience has to re-read that on every slide. This reading of irrelevant information uses up a part of the brain's capacity, which is not available any more for understanding the important contents of the presentation and it needs precious time. In addition this just bothers your audience!

That means, that due to the reasons mentioned above the **footers of slides should contain as little information as possible**. If you show your presentation with a beamer the page numbers are superfluous. They are only helpful, if you show your slides with the overhead projector. But there is no rule without exception: For advertising and prestige purposes and due to juristical reasons many institutes and companies have corporate identity (CI) rules, which define a certain framework for all their presentation slides. This includes showing the name and logo of the company. We did that in our presentation example as well. Copyright notes may be necessary to reserve the company's copyright. But this occurs seldom. Sometimes after filling up the slide with formalities like this there are only 50 % or less of the total slide area available for the actually interesting information. Try to minimize or eliminate this kind of side information as far as possible. A more attentive audience will thank you.

However, **for notes and handouts** you should use **footers** (menu View – Headers and footers, Notes and handouts tab), because e. g. with page numbers you will facilitate the paperwork for your audience!

Creating presentations with Open Office Impress

Under **Linux** you can use Open Office 2.0 Impress for creating presentations. This version of the program is quite similar to Microsoft PowerPoint. In the following you will find the specialties standing out to an experienced Windows user when creating a simple presentation with Impress.

After the start of the program an assistant guides you. Open an **empty presentation**. Leave the option **Select slide background** as it its "<Original>" and at **Slide change** select "No effect". An empty presentation will be opened. On the left you see the slide order, in the middle the working area and on the right the task area with the menu areas Master pages, Layouts, User-defined animation and Slide change. At the beginning the area Layouts is opened.

Inserting a new slide works as in PowerPoint with the menu Insert – Page, unfortunately there is no short key for that. Working in the **structure view** is **nearly identical** as in PowerPoint.

By default new slides are not inserted in the layout Title and Text, but **in the layout last used**. The layout "empty slide" is the default. You can assign a new layout to the current slide by clicking a layout in the working area. Therefore look into the slide are on the left, select a slide from the slide order by clicking it, and then select a layout on the right by clicking it in the task area. This layout is used for the current slide in the working area.

The **font names** are different. Arial corresponds to Helvetica, Times New Roman to New Century Schoolbook and Webdings to Zapf Dingbats. By trend the default settings result in smooth and serious presentations.

Editing text in a predefined **layout area** (here: in a text field) works as usual, but there are nine indentation levels available by default.

A **structure at the left margin for transparency assurance** can be created on the first slide after the title slide and the structure slide as a text field. Yet, the text field cannot be selected as graphic object, copied in one slide and inserted into another slide at exactly the same position. Therefore you have to select the text in the text field, copy it, create a new empty text field on another slide and insert the text there.

The **formula editor** takes getting used to. Unfortunately the menu Insert – Figure does only offer "From file...", the **ClipArts** known from Microsoft Office **are missing**. The functions Print, Save and Export as PDF work fine, if the corresponding peripheral devices are configured under Linux.

If you want to use **another master page**, open the menu for the master pages on the right side in the task area. The master page which is currently in use is displayed with a thicker border. If you click a different master page, it will be used for the whole presentation. To edit the **slide master** and **change** the **basic settings** for font type and size, bullets, indention depth etc., open the menu View – Master – Slide master.

To assign a **user-defined animation** you have to select an object which shall be animated in the working area, e. g. a text field. The whole text should be selected. On the right in the task area click on Add and select an animation, e. g. Appearing. By default the complete text will appear at the same time. To let the list items appear consecutively, click the items one after the other on the right in the task area, and change the option Start from "With previous" to "On click". In the list symbols for the type of animation will appear as in PowerPoint. Assigning **Slide changes** works identically as in PowerPoint.

If you have created **graphic objects** with the functions in the **toolbar Drawing**, you can change their look in the menu Format in the submenus Area and Line. **Graphic objects and text objects as well as layout areas** are **duplicated** with the menu function Edit – Duplicate.

But now let us go back to elaborating your presentation.

5.4.2 Step 4: Summarizing the text and working out the details

Step 4 is the practical realization of the details planned so far. What belongs into the 4 minutes **Introduction**? In general the **presentation target** and the **course of action** in the presentation or lecture. In addition, a demarcation is made, which aspects of the topic the speaker cannot or does not want to deal with. The course of action can best be explained by means of the projected structure slide, which he/she has also often handed-out to the audience (the first present!).

In section 5.3.4 we have already discussed how the structure of a presentation or lecture is build up (in contrast with the structure of the report described in section 2.4.4). **Figure 5-7** shows a structure design plot which will later be used for the example presentation "Burner-integrated suction when welding".

*It is important that you **leave out everything which is unnecessary** on the slides: No "Speaker", no "Presentation of ...", no "Contents", no "Structure", no "Welcome", no "Introduction", no "Closing words", no "Thank you"! This structure shall not represent all the contents of the presentation – **it shall just give a rough orientation** for the audience.*

The audience will quickly get used to this short and dense presentation structure and the speaker can have this structure completely present in his/her mind and always in sight, what makes your presentation appear more complete. The **graphic design of this structure** can vary as much as you like – as long as it is **not overburdened**. The few, a little vague items of the structure also have the advantage, that they do not determine the speaker too much – the audience will check whether all items have been covered in the presentation without mercy.

Structure design plot		20 min. presentation	60 min. lecture
<u>Title:</u>	. . . clear, short, simplifying, but . . . still accurate, memorizable!		
<u>Speaker:</u>	. . . (Title) . . . first, (middle) and last name, . . . Place (city or address of the company)		
<u>Introduction:</u>	. . . informative, inspiring, demarcating!	4 Min.	8 Min.
<u>Main part:</u>	1. clearly structured, 2. not too many sub items, 3. easy to follow, logical, consequent, 4. fascinating, dramaturgy-oriented!	14 Min.	48 Min.
<u>Conclusion:</u>	. . . summarizing, harmonizing, nothing new!	2 Min.	4 Min.

Figure 5-7 Structure design plot for 20 to 60 minutes presentation time

Now this structure design plot will be used for the example "Burner-integrated suction when welding".

> *Burner-integrated suction when welding*
> *Franziska Benz, Hannover*
>
> <u>*Introduction:*</u>   *Health vs. seam quality?*
> <u>*Main part:*</u>   *Suction of welding pollutants –*
> * *burner-integrated*
> * *optimized design*
> * *verified quality*
> <u>*Conclusion:*</u>   *Summary, outlook*

If the author and the speaker are different persons, this must be specified!

The further summarizing of the text shall cover exactly this framework and **all changes to the next item** in the structure shall be **noticeable for the audience**. Items which have been completely covered can be ticked off at the flipchart, the current item in the structure can be emphasized in a structure at the left margin. A presentation made *transparent* like this gives more security to the speaker and to the audience, the speaker is always the "master of the situation".

Working out the details contains the selection of the exactly suited **facts, figures and statements for each item in the structure**. Restrict yourself to the essential information and leave out all excurses, as interesting as they are for you! Parallel with this selection you create (more) visualizations, the text slides and the manuscript (Step 5).

If you have turned back the pages, you have probably noticed the discrepancy between the exact, detailed backbone of the example student research project thesis and the short, nearly superficial structure of the presentation. That is desired – the speaker has planned to show a lot, but she does not show all her cards and thus remains more independent and flexible.

The preparation of the presentation contains the important **Step 5:** Visualization and manuscript, whose special properties will be described in the next section.

5.4.3 Step 5: Visualization and manuscript

Technical presentations and lectures cannot be presented without images (visualizations). This statement is not only true for presentations, but also for the Technical Report, whose visualization is described in detail in subchapter 3.4. However, for presentations there are a few more rules and recommendations, which will now be introduced.

In contrast with readers of a report, who can look at each visualization under individually selected conditions, look at them as long, intensive, and detailed as they like and study the explaining texts, an audience can only consume what is shown by the speaker during the presentation in a **passive mode**. And the audience shall also follow the speech and compute the provided information. **Checklist 5-3** shows the influence factors on the visualization of the presentation and their main properties. The visualization shall be "striking". What effect does a striking design have? Which properties does a good poster have?

- It is appealing.
- It acts fast.
- It transfers simple messages.

Checklist 5-3   Visualization of the presentation

The four main differences between visualizing a report and a presentation are:

- Time factor
  The time provided for looking at the individual visualizations is
  only determined by the speaker, and in most cases it is quite short.

LIVERPOOL JOHN MOORES UNIVERSITY
LEARNING SERVICES

- Reproduction quality
  It is logically lower because of the necessary projector or beamer and projection wall.
- Larger distance
  Compared with the normal reading distance of 30 – 40 cm
  the distance in a presentation is 10 m and more.
- Disturbing influences
  They derive from other people in the audience and room factors like too bright light, glare or obstacles to look at the projection wall, e. g. the speaker or an arm of the projector can trouble your audience.

These conditions require a presentation-suited visualization with the following properties:

- Minimum contents
  Restriction of the message of the image to the absolute minimum, no sentences or even paragraphs to read, no complex illustrations.
- Striking design
  Emphasized, evtl. exaggerated layout with thick lines,
  simplified illustrations and extra large labels.

Your visualizations shall not be similar as advertising for cigarettes or cars, but their design should serve you at least a little bit as an example for the presentation-suited creation of your own visualizations.

Due to the reasons named above visualizations from books, journals and reports cannot be used directly in presentations in most cases, unless (which is an exception) they have already been created in a presentation-suited way!

Visualization tips

There is no "patent medicine" for the successful visualization of presentations, but a general direction to follow, see **Checklist 5-3**.

1. Slimming! Minimizing the message and contents of the image to the essential!
2. Makeup!   Poster-like design and emphasis

The **Slimming** needs objectivity and creativity to find the essential and to formulate short and precise. This is also called didactic reduction, see also **Figure 3-14**. The **Makeup** contains the following aspects:

Heading:	striking, meaningful, correct; contains central message; short, precise spoken language!!
Structure:	logical arrangement of 5-7 elements (words, items, figures)
Reading dynamic:	eye-catcher = starting point: if not in the upper left corner, then you should propose a reading sequence (direction, orientation) emphasized shape, color, size or "cloud"; end point: in the lower right corner, otherwise emphasize it.

| Use of color: | wherever it makes sense, but well-directed! Every color must be explainable! Keep color psychology in mind! (No riot of colors = sensory overload) |
| Font size: | varied between 2 sizes, but never too small! |

☞ To test your visualization think about two aspects:
Does the visualization work in DIN-A4-size from 1,8 m distance
and after 3 – 5 s looking at it?
(Put the visualization onto the floor and look at it when standing upright.)

After these rules and test criteria here is another important tip:

The visualization must not make the presenting person superfluous!

What does that mean?
1. Literal reading of text bores your audience (they can do that themselves).
2. The labels in images should let room for the speaker's explanations!
3. Texts on slides should not be sentences, just keywords!
4. Personally affected things should not appear on slides!

Regarding 1. and 2.: The speech shall be accompanied and emphasized by the visualization, but not replaced. The balance between information provided in images and in the speech must be carefully designed.
Regarding 3.: See the slides in the visualization example! The last slide is an exception, in the summary short mnemonics may be useful.
Regarding 4.: Words, sentences or formulas like "Good Morning, ...", "Any questions?" or "Many thanks for your attendance!" must be said and not shown. These are important means for your personal contact with the audience.

These hints in abridgment shall emphasize and complete the recommendations in subchapter 3.4. In addition, a specialty of the presentation visualization shall be addressed: the animation.

Animation

An *animation* (lat. sensitization, vitalization) we understand all that lives, i. e. all that moves in a visualization during the presentation. In the report there are only static images. The presentation is loosened, emphasized and improved at least by the image changes, but also quite frequently by the animation of parts or elements of an image (if this is not exaggerated).

Two good examples for animations:
1. *List items appear one after another, blink or change their color when they are addressed.*
2. *Complex or time-dependent, discontinuous relations or processes, like the scheme of the production of a motor, the kinematic of a sewing machine or the material flow in a power plant can be displayed clearly (but costly), by emphasizing the current phase with a background color, a frame, strong colors or blinking.*

But: Avoid time-dependent animations, if you have to fight against stage-fright! It is better, if you click each step on your own (wireless mouse)!

Poor animations:
Constant movement on the projection wall, steady stimulus satiation by action and shapes and evtl. additional background sounds like fizzling, humming or scraps of music, which let appear the presentation like a film and make the speaker super-fluous.

Conclusion:
The more possibilities of a modern presentation there are, the more sensitivity of the speaker is needed, to prevent losing the main effect of the presentation – the personal impression and the charisma of the person!

Two hints for the presentation with laptop and beamer:

a) Already today and even more in the future using a laptop and beamer is no at-tribute of quality any more. The fascination from the beginning gives way to customization. What adds up in the end are the technical and personal contents of your presentation again.

b) Presentation equipment may not operate properly. – Always take with you a set of overhead slides to important presentations!

Last but not least here is a small trick, which may help you in a computer-controlled beamer presentation: Let the **presentation program display a timer** with start at 00:00:00 in the lower left corner of the screen, then you have an exact time control in sight, which helps you fight against the time. In PowerPoint you can activate the timer in menu **Screen presentation – "Test new show times"**. At the end of your presentation you should answer the question, whether you want to save the show times with No.

Visualization example
Now our example *"Burner-integrated suction when welding"* will be visual-ized in eleven slides. The basic layout of the slides in our visualization example shall contain the structure at the left margin. The current point of the presentation structure is emphasized with bold print. The emphasis can also be made with a dif-ferent color like red. On the title slide and the structure slide the presentation structure at the left margin and the heading can be left out. The structuring lines can also be left out.

Title slide (Slide 1 of the example presentation)

Structure slide (Slide 2 of the example presentation)

Burner-integrated suction when welding **Health vs. seam quality?** Suction of welding pollutants – - burner-integrated - optimized design - verified quality Summary and outlook	**40 years with welding pollutants?**  **Hannover University** ❘▄❘▪❘

Introduction slide (Slide 3 of the example presentation), "Well-known"

Burner-integrated suction when welding
 	
Health vs. seam quality?
 	
Suction of welding pollutants –
- burner-integrated
- optimized design
- verified quality
 	
Summary and outlook
	Causes of welding pollutants
	
	• Dusts and gases
	• caused by
	  – burning
	  – vaporization
	  – chemical reactions
	• of
	  – part metals
	  – filler metals
	  – welding auxiliary materials
	• alone, with each other and with the atmosphere
	
	Hannover University ❘▄❘▪❘

Causes of welding pollutants (Slide 4 of the example presentation), "Well-known"

Hints on the design of slide 4:
Compact, but clearly many technical terms are explained in the presentation, i. e. the slides should not be totally self-explaining and should not contain complete sentences. If they were, what are you needed for as the speaker?

Hints on the presentation of slide 4 in PowerPoint:
Using "Screen presentation – Predefined animation" you can cause dynamic effects: At first "dusts" and "gases" appears, then you press the space bar or the wireless mouse, "• caused by ..." appears, then "• of ..." and finally "• alone, with each other ...". Using "Screen presentation – Animation preview" you can examine the effect. But be careful! Often less action results in more attendance!

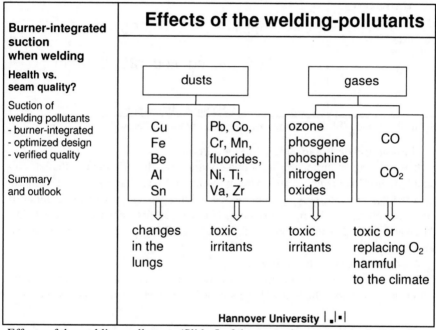

Effects of the welding pollutants (Slide 5 of the example presentation), "New"

Hints on the design of slide 5:
The design is clear and tidy, the text boxes have the same size (different sizes would disturb the esthetic). When explaining the abbreviations of the dusts and gases, you can actively include the audience (those who know the abbreviations are happy; the others learn something). Slide 5 of the example presentation can be omitted or kept as reserve, if time gets narrow.

Hints on the design of slide 5 in PowerPoint:
In PowerPoint this slide is made up with framed text boxes, which are connected with "AutoForms – Connections" in the toolbar "Drawing". Alternatively (within

a table in Word) you can write text into a paragraph, use tabs to position the text and draw rectangles and lines above the text.

Burner-integrated suction when welding Health vs. seam quality? **Suction of welding pollutants** - burner-integrated - optimized design - verified quality Summary and outlook	**Suction methods**	
	Method	**Suction volume**
	• stationary	$800 - 1500 \text{ m}^3/\text{h}$
	• partly stationary	$180 - 400 \text{ m}^3/\text{h}$
	• mobile	$60 - 80 \text{ m}^3/\text{h}$
	• burner-integrated	$25 - 30 \text{ m}^3/\text{h}$
	Hannover University ▎▪▎	

Suction methods (Slide 6 of the example presentation), "New"

Hints on the design of slide 6 in PowerPoint

If you define an animation for a two-column text, the first item in the left column "Method" is displayed first, after pressing the space bar the second to fourth item will appear and then the four suction volume values in the right column. If the right column (= the right text field) is displayed first, change the text fields. Alternative: Use a one-column text field and a tab before the suction volume values.

Another slide (not shown here):

Slide 7 of the example presentation shows the conventional suction methods (hall, workplace, collector which must be positioned by the welder) in coarse sketches.

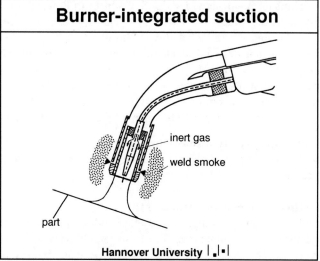

Burner-integrated suction when welding

Health vs. seam quality?

Suction of welding pollutants
- burner-integrated
- optimized design
- verified quality

Summary and outlook

Original suction nozzle (Slide 8 of the example presentation), "New"

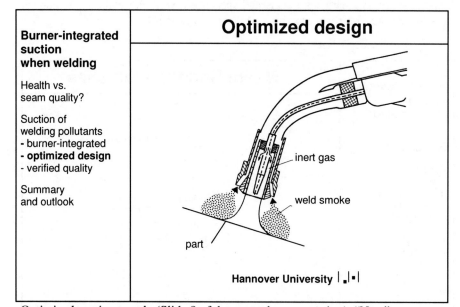

Burner-integrated suction when welding

Health vs. seam quality?

Suction of welding pollutants
- burner-integrated
- optimized design
- verified quality

Summary and outlook

Optimized suction nozzle (Slide 9 of the example presentation), "New"

Hint on the design of slides 8 and 9:
The section drawings of the burner with the original and optimized nozzle cannot be large and clear enough!

Hints on the design of slides 8, 9 and 11 in PowerPoint:
To insert the slides 8, 9 and 11 you can e. g. use the AutoLayout "Title only". Then the images are included with "Insert – Graphics – From File". Both is possible: to include scanned pixel files or pixel or vector files drawn with a graphics program. If you include pixel files, you should not scale them with the handles, but with "Format – Graphics – Size". By entering the same values for x and y direction you can avoid distortions. If there are strange Moiré patterns in the printout, scale your original graphic by multiples of 2 (25 %, 50 %, 200 %, etc.).

Another slide (not shown here):
Slide 10 of the example presentation shows the variation of the suction rings in sketches.

Hints on the design of slide 11:
Slide 11 of the presentation and other slides which are not shown here make up the theoretical, complex part for the experts ("Mazy"). Such images shall be more complex. The experts find their way through many figures, elements and symbols. A much higher information density is a part of the presentation tactic here: The experts study and learn the details and specialties while the majority is more or less overburdened for a not too long while and waits in awe.

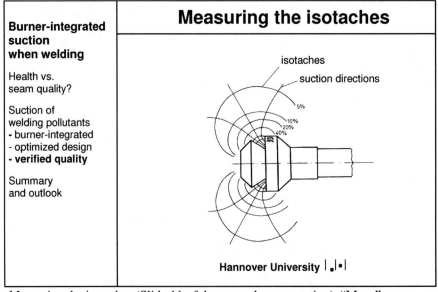

Measuring the isotaches (Slide 11 of the example presentation), "Mazy"

More slides (not shown here):
As already mentioned, 2-3 slides with demanding contents ("Mazy") and 1-2

slides showing the experiment design ("New") are not shown here. The second last slide forms the end of the main part of the presentation and brings new, clear messages to everybody in the audience.

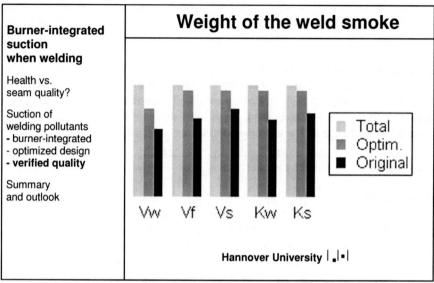

Burner-integrated suction when welding	**Weight of the weld smoke**
Health vs. seam quality? Suction of welding pollutants - burner-integrated - optimized design - **verified quality** Summary and outlook	Vw Vf Vs Kw Ks ■ Total ■ Optim. ■ Original **Hannover University**

Results of weighing the weld smoke
(second last slide of the example presentation), "New"

Burner-integrated suction when welding	**Summary and outlook**
Health vs. seam quality? Suction of welding pollutants - burner-integrated - optimized design - verified quality **Summary and outlook**	• Seam quality is without flaws! • Raised amount of inert gas 150%, for fillet welds 200 %. • Savings in hall equipment and/or workplace equipment. • Burner-integrated suction is easy and safe to handle. • Efficiency is possible. • Welder's health and seam quality are not a contradiction! **Hannover University**

Summary (last slide of the example presentation), "Nothing New!"

Hint on the design of the last slide:
In the summary all profound results or key statements of the presentation should be listed and spoken about without introducing essentially new aspects or informa-

tion. With these points the general understanding of the contents of the presentation shall be assured and a smooth transition to the further in-depth discussion shall be facilitated.

These examples of visualizations show you a small insight from the huge amount of possibilities. All decisions you make are a question of your personal preferences, your software and the conditions of your presentation task.

Presentation manuscript
This will be a short section, because you should not have a **manuscript** and speak out free (best impression). But not everybody succeeds in that, at least at the beginning of the career. Select the form of presentation manuscript you need as a guide for your thoughts. The following options are available, **Checklist 5-4**:

Checklist 5-4   Forms of a presentation manuscript

- **Formulated text**
  This should be avoided, because it is a speech barrier. Exception 1: You read everything literally – that would be a pity for your presentation. Exception 2: You use a formulated text only for opening and closing your presentation, in case of being in poor shape on that day, and for literal citations.

- **Key sentences on DIN A4 paper**
  Mediocre, since there are still reading effort before speaking, paper rustling and constraints to free gestures.

- **Reduced slides and notes on DIN A5 paper**
  In PowerPoint with the function "View – Notes" you can write key sentences or keywords below the slides. Later you can print them in menu "File – Print". At the bottom of this window you should select: "Print: Note pages". These note pages can be reduced with the copier to DIN A5 and then be used as manuscript. This function is also very useful, if you want to speak in a foreign language. You can search relevant vocabulary at home and add them to the slides.

- **Keywords on cards DIN A 6**
  Rather good to good variant, because the cards can be easily exchanged when you create the manuscript and they are easy to handle after a little training. The handling of the cards does not cause much noise and implies only a few constraints to free gestures.

- **Keywords on the visualizations**
  Good to very good variant, if the keywords are written on folding cardboard or plastic bands directly on the slides and therefore not clearly visible to the audience. With that method you can cause the impression, that you are speaking completely "free".

From these options choose the one that is most sympathetic to you, try it out and make it perfect with growing training. The card method DIN A6, see **Figure 5-8**, can be recommended for the beginning. Please keep the following hints and properties in mind:

- Your presentation can be combined, rearranged, made longer or shorter like a mosaique.
- For a new presentation you can use old and new cards (only the front sides!).
- On each card you can note about 5 keywords or figures in as large (block) letters as possible (thick felt pen/bold print).
- The heading contains the structure item and in the upper right corner a consecutive number (written with pencil!).
- In the lower right corner there is the running minute in which the card should be completely finished.
- The cards are punched for filing them in a ring binder.

The following **color system** has proven to be practical:

- **Blue cards:** "direction cards" for welcome, intermediate summaries (redundancy) and closing sentences
- **Red cards:** "must cards" for introductory words, main part and summary as well as all essential facts and figures
- **Yellow cards:** "should cards" for additions to the main part and summary
- **Green cards:** "can cards" as reserve material with additional information and details

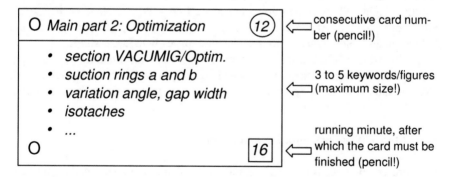

Figure **5-8** Manuscript card

The system works as follows:

- **Blue (direction) and red (must) cards** are essential, they must be processed. Therefore it should be only as many as you can definitely use in the timeframe.
- **Yellow (should) cards** should be processed, if possible, but they can also be omitted (skipped), if you are running short of time.

- **Green (can) cards** are the reserve material, if you have more time than estimated (can happen as well) and if there are special questions.

The colors help to rearrange the structure during the presentation without losing the key message of your presentation (missed topic...).

Presentation handout or documentation
The transitions between manuscript and handout or documentation are floating. Depending on your information target you can hand out the following documents:

- the **presentation,**
- **copied slides** in original size or reduced (in PowerPoint: File – Print – Handout: 2/3/6 Slides per page),
- **Note pages**, if you did not only write a few keywords but more detailed information as addition to the slides (File – Print – Note pages),
- **Copies of various documents**, which have something to do with your presentation topic, like copies of important figures or tables from your Technical Report, which are hard to project and/or
- an **elaborated documentation**.

Providing your audience with **handout pages with 2 slides per page** is the most common type of handout. By default, **PowerPoint leaves a very broad white area around the slides** and does not make optimum use of the paper.

To **change this**, open the menu File – Page layout and select Paper size: User-defined. Then select 60 or 75 cm as paper width and 40 or 50 cm as paper height (width : height = 3 : 2). Then go to the menu File – Print – Handout pages and Slides per page: 2. At the end activate the options "Adopt to paper size" und "Slide frames". Now you can print the handout pages with larger slides.

So far we have created the presentation structure, slides, manuscript, handout and/or documentation. Now we proceed to step 6 with the trial presentation, changes and the necessary preparations, before the actual presentation takes place.

5.4.4 Step 6: Trial presentation and changes

Each presentation needs a trial presentation, if it is important and shall be successful. It should be organized as follows:

- All spoken parts including explanations of the figures and eventual showing films, videos or models/demonstration objects must be executed **completely** and **realistic** (no abbreviations, normal speed of speech and films).
- At least one person in the **audience** to ...
- ... write a **protocol** and take the time of all parts of speech and the total time.

It is best to execute the trial presentation in the later meeting room with its peculiarities, but all other rooms are also helpful. You should then evaluate the protocol honest and objective and rework the presentation, **Checklist 5-5**.

Checklist 5-5   Evaluation of the trial presentation

- Where do I need to leave out what (if it took too long)?
- Where is it necessary to add something for better understanding?
- Which figures are too small, too overloaded, too complex?
- Which keywords, sentences, buzzwords or gags are useful,
  which are embarrassing or irritating? (avoid irony!)
- Which contents do I have to add? Do I need reserve material?

Your audience does not need to be experts to judge a presentation. Sound intellect, interest and honesty are sufficient in most cases, to improve your presentation.

The right consequence of your trial presentation and its protocol is, that you are really willing to change everything what was not convincing. It is only then, that you have the feeling, that you did everything for the success of your presentation. In addition, the feeling of a good preparation makes you more self-confident (less stage fright), because you can show a good product. Even, if there is much to change – do it as good as you can.

Two and more trial presentations are quite usual e. g. for dissertation and habilitation presentations and audition lectures for the job as professors. This is because a radical cutback after the first trial presentation proves to be exaggerated in the second trial presentation and requires new additions etc.

5.4.5 Step 7: Updating the presentation and preparations in the room

The crucial phase shortly before the presentation requires updating the presentation and the preparations in the meeting room.

Updating the presentation
Often the preparation of a presentation runs for quite a while. However, the contents of the presentation shall be really up-to-date, especially for the experts. This requires

- a last check of the contents of the presentation, especially the expert chapters,
- eventually a few looks into the latest editions of scientific journals and the newspaper and
- a talk with the professor, customer, boss or the conference organizers: there could be new conditions!

Preparations in the meeting room
On the day of your presentation you should **be in the meeting room about two hours early,** because there are still four tasks:

- the personal preparation,
- the technical preparation,

258 of the meeting room, and

- the preparation of the meeting room, and
- the preparation of the contents.

The **personal preparation** contains

- Be there early enough (traffic jams and missed trains are no excuse!),
- sleep long enough/be well-rested and fit/in good shape,
- fresh air in the lungs (take a walk),
- correct hairstyle and clothing (mirror!), and
- calm and self-confident charisma (try it).

The **technical preparation** prior to the presentation consists of

- **the image projection** (test it, switch it on/off, adjust the image size, move the projector if necessary, adjust the focus, test the brightness and image sequence; take a spare bulb with you!),
- **the position of the desk** (good for speaking, correct height, not in the field of view; for right-handed persons it is best to have the desk on the right side of the field of view seen in the viewing direction of the audience; depositing rack for manuscript, slides, laptop, pencil, paper, pointer, stop watch),
- **the microphone** (position a fixed microphone, adjust it in speaking direction, watch out for the directional characteristic, evtl. ask for it; fix a clip-on microphone and the radio transmitter to your clothing, switch it on, definitely test the volume!), and
- **test to display models/presentation objects/film projectors** (build them up, try them out and cover them again for a better dramaturgy).

The **preparations in the meeting room** aim at a convenient atmosphere for the audience. They consist of the following steps:

- **The brightness** must be adjusted, so that the images look brilliant, but the audience can still read and write – evtl. ask someone to help you,
- **The air quality** must be guaranteed by ventilation (otherwise the audience might fall asleep),
- Adjust the **temperature** (or let it be done),
- **Clean the whiteboard or blackboard**, remove used paper from flipcharts,
- near the **desk** remove everything that might detract you,
- care for **cleanliness and tidiness** and
- remove **sources of noise** like a loud air conditioning, noise in a neighbored room, close windows against noise from the street etc. (or let it be done)

You might ask: "What are janitors good for?!" That's ok, ask him for help, but during your presentation you are responsible!

The **preparation of the contents** prior to the presentation consists of

- distributing the structure or documentation and
- evtl. showing an introductory slide (like slide 3 of the example presentation) or a slide signaling a break or a welcome slide.

5.4.6 Step 8: Lecture, presentation

This step means actually presenting your slides and keeping all rules and hints in mind. In the next subchapter 5.5 it will be further divided into the phases contact preparations and contacting the audience, creating a relationship with the audience, appropriate pointing, and dealing with intermediate questions.

5.5 Giving the presentation

Finally there is the start. – The nearly unbearable tension can relax. – But how can you reach that? How do you warm up? How do you survive the first critical five minutes? Read and try out the following experiences, rules and tips?

5.5.1 Contact preparations and contacting the audience

Your presentation starts with contact preparations and contacting the audience.

Contact preparations:	*The speaker goes to the desk well-dressed, well-rested, firm, self-confident. <u>She starts the stop watch on the desk</u> (is often forgotten!). Do not use a wristwatch with real time display, because you would often have to compute the speaking time, which disturbs you. The speaker looks around to the audience once, smiling friendly and respectful, looks at the persons in the first row and says:*
Contacting the audience:	*"Doctor Gärtner (president of the German association for welding technology), professor Hauser (chairman of the institute), dear ladies and gentlemen, I can also see the sun shining outdoor! All the more I am happy for turning up in such large number to listen to my presentation!"*

These steps "contact preparations" and "contacting the audience" aim at a **positive optical impression** and thus **acceptance of the person and willingness to listen to the contents of the presentation.**

These two phases should not be underestimated: Those who appear chaotic, arrogant or very shy gain the first prejudice, disadvantage and drawback.

When **greeting your audience** it is important to keep the correct order of **prominent people and the rest of the audience**:

- The "main person" comes first
  (ladies before gentlemen, the district administrator before the mayor etc.).
- Maximum 5-7 people should be individually greeted, otherwise name groups:
  "Dear assemblymen, dear professors".

- "Main person" or "prominent" are also persons which stand out somehow: one lady amongst gentlemen, one man amongst women.
- Politicians should always be mentioned with their names (they want to be recognized!)
- Do not name anybody, who is not present! (awkward, drawback for bad preparation, see section 5.4.5 Updating the presentation!)
- Name titles, but do not exaggerate.

In doubt ask the conference chairman or the secretary to make sure that you greet the prominents with their correct titles and in the correct order – they will be happy to support you. In a negative case the "important" people, who are also often decision makers, are right away from the beginning offended in their vanity and start to revolt in their mind against your presentation. Moreover the audience will take your savoir vivre as a hint for the potential of your company. Do not start any experiments in this phase!

The presentation starts now with an introduction of the speaker, the presentation target and how the topic shall be approached by means of the structure.

Opening: *"My name is Franziska Benz, I am research assistant at the institute for welding technology of Hannover University since one year. My presentation target is to show you the latest findings of burner-integrated suction when welding.*
I want to proceed according to the following structure:"
(Slide 2 appears and is explained without literal reading)

This businesslike opening is always all right, if you do not have a more thrilling idea. If you want to start with a gag ("wow effect"), plan and test it carefully in front of honest, critical test persons. If you introduce yourself, you never name your own title, unless you are in need of it ...

Before the presentation you can literally formulate the greetings and read them literally in case of "start problems". You should *always* write down the names and titles of the important persons to greet and literally read them from your note.

5.5.2 Creating a relationship with the audience

In the human communication, also when presenting the most boring technical contents, it is inevitable to **create** a **positive relationship between speaker and audience**. If there is nothing "going on" between both parties, the presentation does not "come across" well!

This experience is often disregarded by technicians, because they are so convinced of the quality of their subject and qualification, that they do not care enough for "these manners". But "these manners" are more than a container for the technical contents. The manners, the **art of human interaction** in the widest sense, are the **key to introduce technology to people** and to influence the decision makers in a positive way. Therefore engineers should know a little bit of psychology. The more important teams become in our professional life, the more important are key competences like these.

How can you build up such a relationship at the beginning of your presentation? There is no patent medicine, but a recommendation: **Friendliness, open-mindedness and interest for the audience** create sympathy and in response open-mindedness and acceptance for your person and your presentation contents in the audience. Here are some means and methods for creating a positive relationship as examples:

- a successful, well-balanced greeting,
- referring to the common situation ("nice room", "yesterday night we visited ... together" etc.),
- integration of contents from the previous presentation, praise of the previous presentation (if appropriate) and advertising for the next presentation,
- friendly, positive, human introductory sentences,
- offer to ask intermediate questions, organization of breaks, offer to open the discussion,
- distribution of the structure or a first documentation (1st present),
- announcing/promising a more bulky documentation or a little surprise at the end of the presentation or
- anything else positive that comes to your mind ...

In any person, also in yourself, is a child that wants to be hugged – then they are much more open for the message of your presentation.

Thanks and Introduction *Our speaker thanks the head of the institute, her supervising research assistant and some laboratory engineers for the good professional and personal supervision. After these thanks the speaker presents the complete introduction by showing slide 3 and underlining the necessity of health and safety at work when welding. To create concern in the audience she asks the rhetoric question who would like to have his father, brother or sun suffer in such smoke for a whole professional life.*

5.5.3 Appropriate pointing

When explaining the figures you should keep **as much eye contact with your audience as possible**, i. e. you either show with a **pointer** (plastic hand, flat ruler, tapered, not rolling pencil) on an overhead projector or with the **mouse cursor** on the display of the laptop. (Speakers who use a beamer cannot avoid to point with a cursor.) It was the idea of the overhead projector to point while looking to the audience! **Telescopic pointer and laser pointer should stay at home** to avoid to speak "with the projection wall".

If you point on an overhead projector with a pencil, lay it down onto the slide, let it loose and move it to point something else, the audience can see the pointer direction longer, more sharp-edged and better. If you keep the pencil in your hand and tip on the slide, the audience can recognize with pleasure the trembling of your hand (strongly magnified) and can imagine your stage freight very well. The latter is especially true for the often wildly dancing red point of the laser pointer. Now the first intermediate question is asked. What are you doing?

5.5.4 Dealing with intermediate questions

Generally speaking a modern audience does not want to be silenced. Intermediate questions belong to the presentation time and cannot be forbidden rigorously (that seems unsecure, inflexible and dictatorial; hence, you have to plan some time for that!). There are two big categories of questions: real questions and unreal questions.

Real questions can refer to the organization, bad understanding, problems or just be difficult.

- You should answer questions referring to the **organization** (e. g. switching on the room light) and **questions due to bad understanding** (e. g. a word or figure was not understood) **at once, friendly and short**.
- You should check **problem questions**, which require a longer answer, whether you want to answer them at once or after the presentation. **Then you should gain time** by repeating the question for the whole audience. Then you should answer the question short and precise or ask, whether it is **acceptable to answer the question after the presentation** (note the question!).
- **Difficult questions**, which you cannot answer should either not be answered or you just announce an estimation or assumption ("Please do not nail me down to it!"). Pass the question on **to the audience – often someone knows it**.

Unreal questions are no questions, but opinions, self-expressions, objections or pure distortions, e. g. by competitors. Countermeasures:

- recognize that in time (a matter of training), react friendly, but tight,
- depending on the situation "answer", appease or refuse that,
- do not allow yourself to be provoked,
- evtl. let the audience vote and in any case ...
- **stay able to pull the strings!**

Finish   *Our speaker shows the last slide, explains it and says while opening the last slide with her communication data: "Dear ladies and gentlemen, I am happy, that you have listened so curious. This was my first presentation, in the beginning I was very excited; but now I feel fine. I thank you for your attentiveness and would like to answer all your questions now.*

> *Many thanks!" - - - Applause - - - now the speaker hands out her de-*
> *tailed documentation.*

Always set a clear finish and do not let your presentation "melt away"! The audience needs a clear signal for its applause, and this should come from you.

5.6 Review and analysis of the presentation

Is it possible to review such a complex and diverse activity like a presentation? What is more important? The contents or the show, the rhetoric? The opinions of experts will vary here. Nevertheless we want to try to evaluate and review a presentation as appropriate, selective and objective as possible, even if it is only for educational purposes.

First of all, you should answer yourself the questions in **Checklist 5-5** again, which we used to evaluate the trial presentation and compare the answers for the final presentation with the answers for the trial presentation(s). Then we want to evaluate the final presentation a little deeper. The basic idea of the following evaluation scheme, **Checklist 5-6**, is the balanced weighting of all elements of a technical or scientific presentation.

These figures have been selected from experience: they have proven to work well. The weights express, that the rhetoric or the contents alone do not make up the total quality of a presentation. Similar to the analysis of a human person the sum of all properties matters. This does not make the task easy, because you have to care for so many things at the same time. On the other hand this is a chance for every speaker, to emphasize his/her strengths and to compensate weaknesses.

Checklist 5-6   Evaluation of the final presentation (overview)

• Performance (including rhetoric)	30 %	weight
• Contents (amount and level)	30 %	weight
• Organization	20 %	weight
• Impression (subjective)	20 %	weight
	100 %	result

The following scheme would probably look different for other disciplines like humanities. At the same time it is a property list or checklist for the preparation of a presentation. The 25 criteria will now be shortly defined, **Checklist 5-7**:

Checklist 5-7   Evaluation of the final presentation (criteria)

Performance
- Atmosphere: interpersonal and meeting room conditions
- Introduction: contacting the audience, introduction,
                       evtl. "wow effect"
- Rhetoric:
  - Speech flow: velocity, variations, breaks
  - Volume: too calm, nice, too loud
  - Understandability: distinctiveness of the voice, accentuation
  - Mimic: facial expression
  - Gestures: Usage of hand, head and body motions
  - Standing: firmness, posture
  - Eye contact: eye contact and leading the audience
  - Charisma: impression of personality, engagement
                       and persuasive power
- Tension: tension flow, dramaturgy
- Flexibility: reaction after questions, slip-ups, distractions
- Finish: recognizability, harmony, conclusion

Contents
- Amount: too much, well, too few (no reserve material)
- Level: too high, well, too low (bad mixture)

Organization
- Preparation: material selection, effort, planning
- Structure: structure, logical design
- Transparency: clarity, straightforwardness,
                       recognizability of the structure items
- Visualization: visual display of important facts
- Usage of Media: selection and usage of media (blackboard or
                       whiteboard, overhead projector, laptop and beamer,
                       models)
- Redundancy: repetitions and intermediate summaries
- Timing: duration of the presentation parts, keeping the
                       lower/upper time limits

Impression (subjective)
- Learning success: "My gain of knowledge and understanding: ..."
- Identification: "I have adopted the following facts, opinions and
                       arguments: ..."
- Motivation: "My desire to deal with this topic, learn more
                       and evtl. plead for it: ..."

Using **Checklists 5-6 and 5-7** you can calculate as follows:

a) For each of these criteria you can give the following grades.

very good, very often, accurate:	grade 1
good, quite often, rather accurate:	grade 2
average, mean, a little inaccurate:	grade 3
little, seldom, low, rather inaccurate:	grade 4
not acceptable, not visible, very inaccurate:	grade 5

b) Form intermediate sums for each group of criteria (sum 1 to 4).

c) Finally weigh the intermediate sums to get the result.

intermediate sum 1 (performance):	sum 1	$: 13 \cdot 0{,}3 =$
intermediate sum 2 (contents):	sum 2	$: 2 \cdot 0{,}3 =$
intermediate sum 3 (organization):	sum 3	$: 7 \cdot 0{,}2 =$
intermediate sum 4 (impression):	sum 4	$: 3 \cdot 0{,}2 =$
result:		$= \underline{\underline{\quad}}$

Review

This evaluation scheme delivered the following results or weaknesses in more than 800 presentations:

- There should always be a **handout** with the structure lying in front of everybody in the audience!
- A precondition for a serious, successful presentation is **good clothing**.
- At the beginning it is important to greet all prominents and others in the audience correctly and to build up curiosity and tension by announcements, promises or questions. These do not appear by themselves!
- Above every image at the wall there should be the appertaining chapter heading or a heading that can be easily assigned to the current structure item to provide the audience with calming **transparency**.
- The rhetoric is mostly good, but sometimes the speaker **literally reads** everything or at least too much (write only keywords and figures into the manuscript!). Another weakness is a **lack of mimic and gestures** (due to stage freight, too much concentration on the topic and a not very relaxed attitude). These are typical errors of technicians when presenting.
- Often the contents does **not** contain **the three levels** – Well-known, Mazy and New (details, specialties, insider experiences). Then the contents is either too superficial (does not contain expert knowledge) or it assumes too much as "understandable" (contains **too many abbreviations and too much technical jargon**).
- The presentation often contains too much theory at the beginning and **too few examples, descriptions, stories** (better is to first bring practice and descriptive stuff, that create curiosity for the theory!).
- Nearly always the **redundancy** (summaries and short repetitions between the chapters) runs short – it **must** explicitly **be integrated into the manuscript!**

- Often there is no or **too little reserve material** prepared, with which a too short presentation can be inconspicuously expanded to the right amount of time.
- The organization is often too intransparent and has a lack of redundancy and time problems.

Try to keep these very important points in mind and to avoid any weaknesses. But: One step after the other! It needs much training and a few bitter experiences from own presentations, until you are perfect. So, after your presentation you should write down all strengths and weaknesses of your presentation immediately. If you regard these items for your next presentation, you will steadily become better and have more and more success and fun when presenting.

5.7 57 Rhetoric tips from A to Z

On popular demand of our readers here are a few rhetoric tips from A to Z.

Audience: the addressees of your presentation, but also participants and partners, who want to be treated like that and like humans, not like scanners. Otherwise the audience goes on strike and you talk against a wall.

AXLR5 does not mean something to everybody – explain all **abbreviations**, otherwise you will earn questions.

Accentuation (the right one) is inevitable for a vivid presentation style and for understanding, especially for foreign words and proper names (when in doubt, look them up or ask someone).

Arrogance is out of place even for a total expert. It makes you dislikable and disturbs the acceptance of the audience for your message.

Body language via gestures and mimic turns a presentation into an event and experience for the audience, that continues to have an effect.

Breathing should not be done like a singer, but imperceptibly; that means as short and often as possible or "quick, silent and without effort".

Chance: Every rhetorically successful presentation provides you with it – a good critique, a credit, an order, funding or a career step may follow.

Charisma of the speaker: It is based on competency in the field, engagement, sovereignty and an open and friendly communication, see also personality.

Clothing is the first optical impression; it shows your engagement for the audience; it shall be adequate to the situation, the contents and your person, when in doubt preferably traditional and better than usual.

Contact to the audience is the main advantage of the speaker in contrast with a book or video. Design it with care by greeting, eye-contact, questions and personal remarks. To do that, do never use images! ("Many thanks for your attention!" on the last slide is contra-productive).

Controlling himself and the audience is hard for the speaker, but inevitable: time, attention of the audience as well as the quality of speech and images must be constantly verified.

Demagogy may never be the presentation or speech target, so do not present false or unfair contents!

Dialect is human and you can hear it nearly everywhere; it should be neither suppressed nor exaggerated; it shall not disturb the understanding of the audience, not seem ridiculous and not part. Ideal is a "thought high-level language".

Eye-contact is the first bridge to the audience, creating contact, respect and trust; these are the prerequisites to create acceptance in the audience.

Feet are visible in most cases and part of your body language, use them in a natural and inconspicuous way.

Filler words and phonemes are dispensable and do not show discipline of thinking! Instead of beginning a sentence with "Er(r)", "Well, ...", "And ..." and "Actually ..." you should better make a pause.

Foreign words should always be explained or translated (along the way with ease) before someone can knit one's brows, that would just cost seconds and avoids questions. Using foreign words is no guarantee for scientificity.

Formulation is the most important means to achieve understanding. Form clear, short sentences in personal colloquial language and avoid a too scientific or officialese speech style.

Gestures are the support of your presentation contents by means of body language. Holding your hands in the height of the bellybutton is the best and easiest initial position for optically emphasizing important (not all) items.

Greeting is the first verbal contact with the audience. It shall be planned carefully (prominents ...) and easily and winningly build the first bridge.

Hands are a problem only at the beginning. They should be always visible and as often as possible without holding things (cards, pointer, mouse) to be able to make gestures.

Humor is the spice of your speech; it should never miss completely (also humorous self-criticism), always be well-tested and never be embarrassing or rassistic.

Impression is the image you create in your audience. To achieve a good one you may exaggerate from time to time, behave different or act a little.

Influencing is one aim of good rhetoric. The audience shall be motivated to understand, to positively decide (funding, buying ...), to personal trust and to appreciate the speaker.

Inhibitions are minimized by excellent preparation (trial presentations) and best physical fitness (sleep, clothing, punctuality).

Intelligence is not measured by your scientificity, but by your flexibility, quick-wittedness and art of human interaction.

Intermediate questions should be answered friendly, completely and quickly to cost less time.

Key and foreign words: They should be spoken out very, very clearly and embraced in speech pauses.

Listening to yourself is the recipe for unsuspiciously checking your speech, the clarity, volume and speed: From time to time you should listen to yourself while you are speaking to minimize problems.

Mimic is the language of your facial features; it shall be vivid and natural, i. e. change it from friendly to neutral or serious from time too time, depending on the presentation contents.

Modesty is the opposite of arrogance; but too much of it can be interpreted as shyness (uncertainty). Ideally you are modestly self-confident ("I know much, but not everything.").

Pauses are the best of work. Also in a presentation moments of silence are appropriate: pauses of thinking, pauses to emphasize something, pauses to let something continue to have an effect, but do not make too long pauses in moments of uncertainty or blackout (and never use filler phonemes like er(r), well, and, actually ...).

Personality is your main capital for the success of a good technical presentation. It is not only important, what you know, but also who you are or what the impression of your audience is, see also charisma.

Persuading by speech and supporting gestures and mimic is the result of a good presentation – people must be able to believe, what you are saying.

Pitch and speech intensity should vary during your presentation to avoid monotony.

Pitch of the sentence structure announces a comma in the sentence with higher pitch and a period always with accentuated and decisively lower pitch. Especially a good closing sentence ends in totally low pitch.

Point is the peak of humor; it should be really witty, appropriate and tested, otherwise it is likely to be a disadvantage.

Pointing should never be done against the wall, but while keeping control over the audience with a tapered ruler on the projector or with the mouse on the screen of the laptop! – Point calmly and longer, because not everybody in the audience looks to the front in the same second!

Posture is the optical means to signal security, decisiveness and modesty, i. e. stand upright and tense, but still vivid and always with your front towards the audience.

Questions are the icing on the cake, but they can also freeze your presentation (see 5.5.4). Take it as a racy challenge and prepare necessary answers in advance.

Repetition of important contents, figures and facts supports understanding and remembering.

Review is clever to learn from every failure and success; this is the only way to improvement and routine.

Sayings can be integrated very well, if they fit to the topic. Especially you can tell (or show) the first half of a saying at the beginning of the presentation and the second half at the end to create tension.

Sentence structure should be simple and easy to understand, should not contain nested sentences, see formulation.

Slip-ups can happen, anticipate them! (slip of the tongue, malfunction of the projection technology, mismatch of images ...). Do not exaggerate small slip-ups, but address bigger ones and beg for apology.

Smiling should spice the whole presentation where it fits without appearing too sweet. Technical and serious facts come across better with a friendly face. When there are disturbances, you should always stay friendly (= sympathetic), even if you are very, very upset.

Speech style: It should be oriented towards the middle of the society – there are the decision-makers!

Spoken language does not mean small-talk or chitchat, not the speech style of a user's manual, a law or a government declaration, but a selected, generally understandable everyday speech.

Stage-freight, see inhibitions!

Standing shows the condition of the speaker. You should stand in a natural position, neither stark and stiff nor changing all the time. Then it disturbs the least from your presentation.

Studying all speeches and presentations in the public (TV talk masters, politicians, experts ...) helps to improve yourself. And you should always study your audience for their reactions.

Tension within a presentation – if not planned into the contents – can also be created with your voice, mimic and body language.

Time pressure is the strongest enemy of the speaker. The only precaution are a stop watch and a lot of training to defeat this enemy.

Training is more important than eloquence; frequent presentations and speeches with honest feedback of an audience create experience and routine.

Trust creation is an important target of your presentation – trust into the contents, your person and your company.

Velocity of the speech should be adopted to the contents and its importance; when in doubt speak slower than usual and check by listening to yourself!

Volume (of your speech) depends on the size of the meeting room and the number of people in the audience; it should preferably be louder, but vary from time to time.

6 Summary

Now all details, rules and working procedures relevant for writing and presenting Technical Reports have been introduced in detail. Our network plan for creating Technical Reports has been processed from accepting and analyzing the task to presenting or distributing the final report.

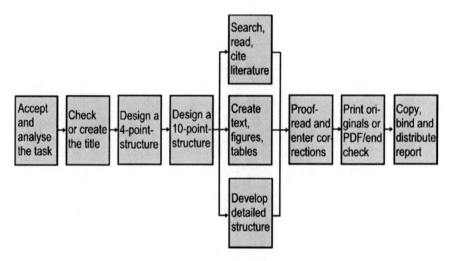

During all work steps the following basic rule needs to be considered. The creator of the Technical Report must always ask at first, whether there are rules issued by the customer or already existing within the own institution, how Technical Reports must be written and designed. When using this book, please keep in mind: Already existing rules (standards of the department or professor or company or customer) must be followed prior to the tips and rules provided in this book.

If such institutional standards do not exist or do not cover all details, you should use the hints and suggestions in this book. The consequent application of the information and working procedures described in this book will probably improve the quality of your future Technical Reports as compared with the quality of your previous ones. Therefore we want to close this book wish good wishes:

☞ *We wish you that your future Technical Reports and their presentation will contribute to your personal success!*

L. Hering, H. Hering, *How to Write Technical Reports*,
DOI 10.1007/978-3-540-69929-3_6, © Springer-Verlag Berlin Heidelberg 2010

7 References

Books, articles etc.

Ammelburg, G.: Rhetorik für den Ingenieur. 5. Aufl. Düsseldorf: VDI-Verlag, 1991

Baker, W. H.: How To Produce and Communicate Structured Text. In: Technical Communication. 41 (1994), p. 456-466

Bargel, H.-J.; Schulze, G.: Werkstoffkunde. 7. Aufl. Berlin: Springer, 2000

Beitz, W., Grote, K.-H. (Hrsg.): Dubbel – Taschenbuch für den Maschinenbau. 21. Aufl. Berlin (u.a.): Springer, 2005

Böttcher, P.: Böttcher/Forberg, Technisches Zeichnen. Hrsg. v. DIN – Deutsches Institut für Normung, 23. Aufl. Stuttgart: Teubner, 1998

Brehler, R: Modernes Redetraining. Niedernhausen/TS: Falken, 1995

Decker, K.-H.: Maschinenelemente: Tabellen und Diagramme. 15. Aufl. München: Hauser, 2000 (außerdem ist ein Aufgabenbuch erhältlich)

Fritz, A. H.; Schulze, G.: Fertigungstechnik. 5. Aufl. Berlin: Springer, 2001

Grünig, C.; Mielke, G.: Präsentieren und überzeugen. Planegg/München: Haufe, 2004

Hartmann, M., Ulbrich, B., Jacobs-Strack, D.: Gekonnt vortragen und präsentieren. Weinheim: Beltz Verlag, 2004

Hering, H.: Verbesserung des Arbeitsschutzes beim Schweißen durch Einsatz brennerintegrierter Absaugdüsen: Effektivität und Qualitätssicherung. Große Studienarbeit, betreut vom Institut für Fabrikanlagen der Universität Hannover und dem Heinz-Piest-Institut für Handwerkstechnik an der Universität Hannover, 1987

Hering, H.: Berufsanforderungen und Berufsausbildung Technischer Redakteure: Verständlich schreiben im Spannungsfeld von Technik und Kommunikation. Dissertation, Universität Klagenfurt, 1993

Hering, L.: Computergestützte Werkstoffwahl in der Konstruktionsausbildung: CAMS in Design Education. Dissertation, Universität Klagenfurt, 1990

Hering, L; Hering, H.; Kurmeyer, U.: EDV für Einsteiger. 2.Aufl. Hemmingen, 1995

Hering,L.; Hering, H.; Köhler, N.: Der TEXTdesigner: Computergestützte Analyse und Optimierung der Verständlichkeit von Sachtexten aller Art. Computerprogramm und Handbuch. Hemmingen, 1994

Herrmann, P.: Reden wie ein Profi. München: Orbis, 1991

Hermann, U.; Götze, L.: Die neue deutsche Rechtschreibung. München: Bertelsmann Lexikon Verlag in Lizenz des Lexikographischen Instituts, 1999

L. Hering, H. Hering, *How to Write Technical Reports*,
DOI 10.1007/978-3-540-69929-3, © Springer-Verlag Berlin Heidelberg 2010

Hoischen, H.: Technisches Zeichnen: Grundlagen, Normen, Beispiele,
   darstellende Geometrie. 30. Aufl. Berlin: Scriptor, 2005

Holzbaur, U; Holzbaur, M.: Die wissenschaftliche Arbeit. München: Hanser,
   1998

Horn, J.: Urheberrecht beim Einsatz neuer Medien in der Hochschullehre.
   Oldenburg: OLWIR Verlag, 2007

Ilzhöfer, V.: Patent-, Marken- und Urheberrecht – Leitfaden für Ausbildung und
   Praxis. 6. Aufl. München: Vahlen, 2005

Klein, M.: Einführung in die DIN-Normen. hrsg. vom DIN, bearb. v. K. G. Krieg
   14. Aufl. Wiesbaden: B. G. Teubner und Berlin: Beuth, 2008

Labisch, S.; Weber, Chr.: Technisches Zeichnen: Intensiv und effektiv lernen und
   üben. 3. Aufl. Wiesbaden: Vieweg, 2007

Marks, H.E.: Der technische Bericht: Ein Leitfaden zum Abfassen von
   Fachaufsätzen sowie zum Vorbereiten von Vorträgen. 2. Aufl. Düsseldorf:
   VDI-Verlag, 1975

Melezinek, A.: Unterrichtstechnologie. Wien, New York: Springer, 1982

Melezinek, A.: Ingenieurpädagogik – Praxis der Vermittlung technischen Wissens.
   4. Aufl. Wien: Springer, 1999

N. N.: Intensivkurs Neue Rechtschreibung. Köln: Serges Medien, 1998

Nordemann, W.; Vinck. K.; Hertin, P.W.: Urheberrecht: Kommentar zum
   Urheberrechtsgesetz und zum Urheberrechtswahrnehmungsgesetz. 8. Aufl.
   Stuttgart: Kohlhammer, 1994

Rehbinder, M; Hubmann, H.: Urheberrecht. 14. Aufl. München: Beck, 2005

Reichert, G. W.: Kompendium für Technische Anleitungen. 6. Aufl. Leinfelden-
   Echterdingen: Konradin, 1989

Reichert, G. W.: Kompendium für Technische Dokumentationen. 2. Aufl.
   Leinfelden-Echterdingen: Konradin, 1993

Roloff, H., Matek, R.: Maschinenelemente. 18. Aufl. Wiesbaden: Vieweg, 2007

Scholze-Stubenrecht, W.; Wermke, M.: Der Duden: Das Standardwerk zur
   deutschen Sprache. Hrsg. v. wissenschaftlichen Rat der Dudenredaktion,
   Band 1: Die deutsche Rechtschreibung. Hrsg. v. der Dudenredaktion auf der
   Grundlage der neuen amtlichen Rechtschreibregeln, 23. Aufl. Mannheim:
   Dudenverlag, 2004

Seifert, J. W.: Visualisieren – Präsentieren – Moderieren. 21. Aufl.
   Offenbach: Gabal, 2004

Theisen, M.: Wissenschaftliches Arbeiten. 12. Aufl. München: Vahlen, 2005

Thiele, A.: Überzeugend Präsentieren. Düsseldorf: VDI-Verlag, 1991

Computer literature

Engel, R.: Microsoft Outlook 2000 auf einen Blick. Unterschleißheim: Microsoft Press, 1999

Feig, M.: UNIX von Anfang an. Frankfurt: Fischer, 1993

Gretschmann, M.; Zankl, M.: Das große Buch PDF mit Acrobat & Co. Düsseldorf: Data Becker, 2004

Herdt: Seminarunterlagen zum Erlernen der Programmbedienung der gängigen Office-Programme, die sich z. T. auch für das Selbststudium eignen, sind erhältlich bei www.herdt.de.

Holland, H.-J. und Bernhardt, J.: Excel für Techniker und Ingenieure: eine grundlegende Einführung am Beispiel technischer Problemstellungen. 3. Aufl. Braunschweig, Wiesbaden: Vieweg, 1998.

Hütter, H. und Degener, M.: Praxishandbuch PowerPoint-Präsentation. Wiesbaden: Gabler, 2003

Kersken, S.: Praxiswissen Flash 8. 2. Aufl. Köln: O'Reilly, 2006 (inkl. CD-ROM)

Kommer, I.; Reinke, H.: Mind Mapping am PC: für Präsentationen, Vorträge, Selbstmanagement mit MindManager 4.0. 2. Aufl. München: Hanser, 2001

Koch, St.: Java Script: Einführung, Programmierung, Referenz. 2. Aufl. Heidelberg: dpunkt Verlag, 1999

Thalmayr, D.: Umsatteln auf Linux. Köln, O'Reilly, 2005 (inkl. DVD mit SUSE 9.3)

Münz, St. und Nefzger, W.: HTML Handbuch. Studienausgabe: Überarbeitete und aktualisierte Neuauflage des Standardwerks. Poing: Franzis, 2005 (inkl. CD-ROM)

Niedermair, E.: LATEX Das Praxishandbuch. 2. Auflage. Poing: Franzis, 2005 (inkl. DVD 9 TEX-Collection)

Ravens, T.: Wissenschaftlich mit Word arbeiten: von Word 2000 bis Word 2003. 2. Aufl., München: Pearson Studium, 2004

Siegel, D.: Das Geheimnis erfolgreicher Websites: Business, Budget, Manpower, Lizenzen, Design. München: Markt&Technik, 1999, Doppelbandausgabe von www.zweitausendundeins.de

Siegel, D.: Web-Site Design: Killer-Websites der 3. Generation. Frankfurt: Zweitausendundeins, 1999

von Wilmsdorff, C.: Praxishandbuch Word, Wiesbaden: Gabler, 2003

Standards, guidelines etc.

German standards are listed with German references, British and international standards are listed with English references. If the references contain an entry „mehrere Teile (oder Blätter)" or „several parts (or sheets)" the standard consists of at least two parts (or sheets), so that it is not possible to list the year of publiccation, because the part (or sheets) of the standard or guideline have bben published in different years.

DIN, Deutsches Institut für Normung (Hrsg.): Berlin: Beuth

DIN	108	Diaprojektoren und Diapositive, mehrere Teile
DIN	199	Technische Produktdokumentation – CAD-Modelle, Zeichnungen und Stücklisten, mehrere Teile, u. a. Teil 1: Begriffe, 03/02
DIN	406	Technisches Zeichnen – Maßeintragung, Teil 11: Grundlagen der Anwendung, 12/92
DIN	461	Graphische Darstellungen in Koordinatensystemen, 03/73
DIN	616	Wälzlager, Maßpläne, 06/00 <Anm. d. Verf.: Anschlußmaße von Wälzlagern>
DIN	623-2	Wälzlager; Grundlagen; Teil2: Zeichnerische Darstellung von Wälzlagern, 06/00
DIN	824	Technische Zeichnungen – Faltung auf Ablageformat, 03/81
DIN	1301	Einheiten, mehrere Teile, u. a. Teil 1: Einheitennamen, Einheitenzeichen, 10/02
DIN	1302	Allgemeine mathematische Zeichen und Begriffe, 12/99
DIN	1303	Vektoren, Matrizen, Tensoren – Zeichen und Begriffe, 03/87
DIN	1304	Formelzeichen, mehrere Teile, u.a. Teil 1: Allgemeine Formelzeichen, 03/94
DIN	1338	Formelschreibweise und Formelsatz, 08/96
DIN	1421	Gliederung und Benummerung in Texten, 01/83
DIN	1422	Veröffentlichungen aus Wissenschaft, Technik, Wirtschaft und Verwaltung, Teil 1: Gestaltung von Manuskripten und Typoskripten, 02/83, Teil 2: Gestaltung von Reinschriften für reprografische Verfahren, 04/84, Teil 3: Typografische Gestaltung, 04/84, Teil 4: Gestaltung von Forschungsberichten, 08/86
DIN	1426	Inhaltsangaben von Dokumenten; Kurzreferate, Literaturberichte, 10/88
DIN	1505	Titelangaben von Dokumenten, mehrere Teile, u. a. Teil 1: Titelaufnahme von Schrifttum, 05/84 und Teil 2: Zitierregeln, 01/84
DIN	5007	Ordnen von Schriftzeichenfolgen (ABC-Regeln), 08/05
DIN	5008	Schreib- und Gestaltungsregeln für die Textverarbeitung, 05/05
DIN	5473	Logik und Mengenlehre – Zeichen und Begriffe 07/92
DIN	5483	Zeitabhängige Größen, mehrere Teile

DIN 6774-4   Technisches Zeichnen; Ausführungsregeln; Gezeichnete
                  Vorlagen für Druckzwecke, Teil 4, 04/82
DIN   6780   Technische Zeichnungen; Vereinfachte Darstellung und
                  Bemaßung von Löchern, 10/00
DIN 16 511   Korrekturzeichen, 01/66
DIN 19 045   Lehr- und Heimprojektion für Steh- und Laufbild, mehrere Teile
DIN 30 600   Grafische Symbole – Registrierung – Bezeichnung, 11/85,
                  sehr viele Beiblätter mit nach Technikbereichen sortierten,
                  genormten Bildzeichen
DIN 31 001   Sicherheitsgerechtes Gestalten technischer Erzeugnisse;
                  Schutzeinrichtungen; Begriffe, Sicherheitsabstände für
                  Erwachsene und Kinder 04/83
DIN 31 051   Grundlagen der Instandhaltung (Basics of maintenance), 06/03
DIN 31 623   Teil 1, Indexierung zur inhaltlichen Erschließung von
                  Dokumenten; Begriffe, Grundlagen, 09/88
DIN 32 520   Grafische Symbole für die Schweißtechnik, mehrere Teile
DIN 32 541   Betreiben von Maschinen und vergleichbaren technischen
                  Arbeitsmitteln – Begriffe für Tätigkeiten (Running machines
                  and comparable teechnical devices – terms for work steps),
                  05/77
DIN 32 830   Grafische Symbole – Gestaltungsregeln für grafische Symbole
                  an Einrichtungen, Teil 1 aus 01/92 und weitere Teile
DIN 55 301   Gestaltung statistischer Tabellen, 09/78
DIN 66 001   Informationsverarbeitung, Sinnbilder und ihre Anwendung,
                  12/83
DIN 66 261   Sinnbilder für Struktogramme, 11/85

DIN EN 62079   Erstellen von Anleitungen; Gliederung, Inhalt und Darstellung,
                  Preparation of Instructions – Structuring, Content and
                  Presentation (IEC 62079:2001)

DIN ISO   128   Technische Zeichnungen; Allgemeine Grundlagen der
                  Darstellung, mehrere Teile, u. a. Teil 20: Linien, Grundregeln,
                  12/02, Teil 50: Grundregeln für Flächen in Schnitten und
                  Schnittansichten, 05/02
DIN ISO   1101   Beiblatt 1: Technische Zeichnungen; Form- und
                  Lagetolerierung, Tolerierte Eigenschaften und Symbole,
                  Zeichnungseintragungen, Kurzfassung 11/92
DIN ISO   2768   Allgemeintoleranzen ... ohne einzelne Toleranzeintragung,
                  mehrere Teile aus 1991
DIN ISO   5456   Technische Zeichnungen; Projektionsmethoden,
                  mehrere Teile aus 1998

VDI, Verein Deutscher Ingenieure (Hrsg.): Düsseldorf, Berlin: Beuth

VDI 2222-2225 Konstruktionsmethodik, mehrere Blätter

VDI   4500   Technische Dokumentation – Benutzerinformation, Blatt 1:
                   Begriffsdefinitionen und rechtliche Grundlagen, 12/04,
                   Blatt 2: Organisieren und Verwalten, 09/05,
                   Blatt 3: Empfehlung für die Darstellung und
                   Verteilung elektronischer Ersatzteilinformationen, 12/01

tekom (Hrsg.): Stuttgart, info@tekom.de

Technische Dokumentation beurteilen. tekom (Hrsg.), Stuttgart, 1992
Richtlinie zur Erstellung von Sicherheitshinweisen in Betriebsanleitungen.
tekom (Hrsg.), Stuttgart, 2005

ISO, International Organisation for Standardization (Hrsg.)

ISO 4:1997   Information and documentation – Rules for the abbreviation of
                   title words and titles of publications
ISO 8:1977   Documentation – Presentation of periodicals
ISO 18:1981   Documentation – Contents list of periodicals
ISO 690:1987   Documentation – Bibliographic references – Content, form,
                   and structure
ISO 832:1994   Information and documentation – Bibliographic description and
                   references – Rules for the abbreviation of bibliographic terms
ISO 999:1996   Information and documentation – Guidelines for the content,
                   organization and presentation of indexes
ISO 1086:1991   Information and documentation – Title leaves of books
ISO 2014###
ISO 2145:1978   Documentation – Numbering of divisions and subdivisions in
                   written documents
ISO 2384:1977   Documentation – Presentation of translations
ISO 3166###
ISO 5776:1983   Graphic technology – Symbols for text correction
ISO 5966:1982   Documentation – Presentation of scientific and technical reports
                   (withdrawn in 2000)
ISO 6357:1985   Documentation – Spine titles on books and other publications
ISO 7144:1986   Documentation – Presentation of theses and similar documents
ISO 7275:1985   Documentation – Presentation of title information of series
ISO 11800:1998   Information and documentation – Requirements for binding
                   materials and methods used in the manufacture of books
ISO 14416:2003   Information and documentation – Requirements for binding of
                   books, periodicals, serials and other paper documents for archive
                   and library use – Methods and materials

BSI, British Standards Institution:London

BS   1629   Recommendations for references to published materials, 1989

A Lists of figures, tables and checklists

A.1 Figures

A.2 *Tables*

A.3 Checklists

B Glossary – terms of printing technology

In the following we will shortly define important terms in the field of printing technology, which might help you to design your Technical Reports and when you are in contact with copy-chops, computer stores, printers or journal and book publishers.

A

Acrobat Reader is a free-of-charge viewer from the company Adobe to read PDF files, which you can download at www.adobe.com/de/products/acrobat/readstep2.html.

An **algorithm** is a calculation scheme. The algorithm contains the logic, the mere „intelligence" of a computer program or function.

The **ANSI** (American National Standards Institute) deals with standardization and has e. g. developed the → ASCII Code.

The **ASCII** Code (American Standard Code for Information Interchange) is a standard to store letters, figures and symbols as decimal number in Byte (in txt files). It can only code the European languages.

B

BW is sometimes used as an abbreviation for black and white.

A **Blog** (Web+Log, internet+notes) is a digital diary or topic page published in the internet.

Bookmarks are used in the internet browser (favorite internet addresses) and in PDF files (navigation through the document on the left side).

Bullets mark the different items in unstructured lists. People either use the common symbols •, •, -, –, — etc. or more "pictorial" symbols from fonts like Symbol, Webdings, Wingdings, Zapf Dingbats etc.

C

A **cache** is a fast intermediate storage for web pages, it contains a copy of already downloaded data on your computer for reuse. If there are display problems, it often helps to empty this storage (Special – Internet options – Delete browser history).

In word-processing, → DTP and spreadsheet programs **cell** is the name of a field in a table, that the user can fill with data.

CMYK stands for the colors cyan (turquoise), magenta (pink), yellow and black. Printers need all color information based on this color system for 4-color-printing.

During **compression** the file size is reduced by a packing program. Moreover, many single files can be combined to one easy-to-handle archive file. In this process a file with the extension *.zip is created. Due to the compression more data can be stored on the storage device. Sending e-mails with a compressed file in the attachment is much faster than without compression.

Condensed is a → font attribute (opposite of → expanded).

Consistency in Technical Reports means, that equal tasks regarding spelling, punctuation and typography are performed always in the same way throughout the whole report.

Correction symbols are used during proof-reading and are standardized in ISO 5776.

cpi (**c**haracters **p**er **i**nch). 10 and 12 cpi are common spacings for fonts with fixed spacing (typewriter or fonts like Courier, Letter Gothic etc.).

A (Web-)**Crawler** (also Spider or Robot) searches the internet, registers URLs, collects keywords and follows links. It mainly helps search engines to find new pages in the internet and to present appropriate hits upon search requests. RSS feed services use crawlers to automatically create news.

D

Decompression or unpacking is the opposite of → compression.

Defense is the oral exam of a doctorate thesis or dissertation. At some universities this is also called rigorosum, disputation, defensio, viva voce.

dpi (**d**ots **p**er **i**nch). 300 and 600 dpi are typical resolutions for laser and ink jet printers. 300 dpi is the minimum resolution of images, if they shall be printed in a journal or book.

Document part is the generic term for all parts of a document independent of their hierarchy level in the structure. It contains chapters, subchapters, sections, subsections etc.

Digital Object Identifier **DOI** for electronic documents
Digital objects in the internet can be found under the address, where they are put on an internet server (URL/URI), but internet addresses are changed from time to time. Therefore the DOI (digital object identification) system was founded by the International DOI Foundation (IDF) to identify the documents themselves. The objects get a number and a server looks in his data base, where the objects are currently available. Example: The object with DOI:10.1007/s003390201377 is found, if you go to the DOI Resolver and enter the DOI number into the window of the server or you enter directly the URI http://dx.doi.org/10.1007/s003390201377 into your browser (i. e. http://dx.doi.org/ and DOI number). You will then be redirected to the server, where the document is available or get a list of links.

DTP (**D**esktop **P**ublishing) is preparing documents at your desk ready to be printed by means of suited DTP programs which can mix text and graphics. Creating two and more column pages and the positioning of figures can be controlled more precisely than with a word-processing program and for later printing → CMYK color separations can be created. Quark Express and PageMaker are well-known page-oriented DTP programs, with Frame-Maker you can create even very thick documents or books very efficiently.

DTV (**D**idactic **T**ypographic **V**isualization) according to REICHERT is visualizing with „text images" (pictorial and tabular re-arrangement of text). Here unstructured lists, boxes, arrows and similar typographic means are used. DTV also makes clear the logical and evtl. hierarchical dependence of text blocks by means of targeted usage of lines, which mostly run vertical and horizontal only. DTV fills the gap between conventional continuous text and graphic visualization and allows easy creation of the text images, interested reading and spontaneous understanding of the relevant facts.

E

In an **evaluation table** different evaluation criteria are compared, which cannot be directly compared as figures. The evaluation criteria get single points assigned. The points are multiplied with weighting factors and these results are summed up over all evaluation criteria. This provides the total points of each (concept) variant. Evaluation tables are often used to compare different concept variants in a design or different locations for planning new company premises or plants.

Expanded is a → font attribute (opposite: → condensed). For fixed spacing fonts you have to write a space character after every letter or figure. If there is a space in normal writing it becomes three spaces in expanded writing. For proportional fonts you can define by how many typographic points the characters shall be expanded.

F

A **figure table** is a table, that (predominantly) contains figures.

A **fixed spacing** font is a font with a fixed character spacing (Courier etc.). In a fixed spacing font the distance from the middle of one letter to the middle of the next letter is constant. Therefore there is some white space on the right and left side of narrow characters (opposite: → proportional font).

Font attributes define the typesetting of the characters and words in a certain font. Font attributes are for example, **bold**, *italic*, underlining: simple, double and dotted, but also crossed-out, superscript, subscript, e x p a n d e d, SMALL CAPS and CAPITAL LETTERS.

G

The word **glossary** is derived from the Greek γλωττα = tongue, language. It is the international name for an alphabetically sorted list of technical terms together with definitions of these terms.

H

Half tone image is an image with smooth transitions of the gray or color values.

The **header** of a table is the top row (line). It contains the generic terms of the entries in the appertaining column. Therefore, it is often accentuated by typographic means (e. g. with a double line or with a gray background).

HTML (Hypertext Markup Language) is the page description language used in the internet and in an intranet. Every reader sees your Technical Report as his browser interprets the commands. The line and page break, you have intended, gets lost.

A **hyperlink** is connection, a reference or a goto command. If you click on the hyperlink, you go to another position within the same file or a program is started and the file or internet address named in the hyperlink is displayed.

I

An **icon** is a small pictorial symbol on the desktop or in a computer program. If you click on the icon, the computer runs a function. That is more convenient than using the menus.

An **inch** is a unit of length. 1 inch = 25,4 mm = 2,54 cm. The resolution of images, printers, copiers and scanners is specified in → **dpi** (dots per inch).

The word **index** is Latin and means forefinger, pointer, overview, title and table of contents. It is the international name for the list of keywords with page numbers to quickly find the relevant text passages.

The **introductory column** of a table is the first column from the left. It contains the generic terms of the entries in the rows. Therefore, it is often accentuated by typographic means (e. g. with a double line or with a gray background).

International Standard Book Number **ISBN** for books (monographs)

International Standard Serial Number **ISSN** (for journals and serial publications)

L

The term **layout** summarizes all measures to influence the appearance of information on the paper. This includes the document or page layout resp. (for example defining the page margins, usage of a page header) and the definition of paragraph and character formats: selection of the font type and size for document part headings, text, figure subheadings, table headings, indentions, accentuations (italic, bold, underlined), usage of bullets in unstructured lists as well as the definition, how labels in images and tables shall be designed.

Library of Congress Control Number **LCCN** (formerly Library of Congress Catalog Card Number) for publications which are registered by the American National Library

Leading dots between document part heading and page number in the table of contents help the eye to hold the line and thus lead the eye to the page number. They are also used in all other lists and indexes with page numbers.

A **legend** is a bulky explanation for tables and figures, which is delivered to the reader in addition to the figure subtitle. A legend is always located *below* the table or figure. Sometimes it appears in a box. In English the term legend is also used as a synonym for figure subtitle and used in opposition to a figure caption (above the figure), which we call figure title in this book to distinguish it from the legend.

The word processor program automatically starts a new **line break**, e. g. if the page margins are changed. A line break before the end of the line can also be manually created by pressing the Shift+↵ keys. (If the text is left-justified and right-justified, you have to press Tab and then Shift+↵.) This is useful in multi-line headings of any kind and lists to order the information into logical units.

lpi (lines per inch). 6, 4 and 3 lpi means line spacing 1, 1½ and 2. This specification of the line spacing is from the type writer and matrix printer age, today it is not so usual any more.

M

A **macro** is an abbreviated command in a computer program, which substitutes entering several other commands (or characters in a word-processing program).

Majuscules (capital letters) is a → font attribute. To write text, only capital letters are used. Text in majuscules is much harder to read than text with capital *and* small letters.

A **matrix** is an arrangement of information in rows and columns.

The **morphological box** is a central element in design methodology. In the morphological box the sub functions of a design and the found design solutions for the sub functions are arranged in a matrix-shape. By combining one solution of each sub function a complete design concept is derived.

Multimedia is the combination of text, tables, images, sound and video sequences (including computer animations) to a new form of information display. If a person perceives such information, several senses are addressed at the same time. This improves the amount of learning and remembering.

N

The **non-breaking space** (NBSP) is e. g. used between the components of a multi-part abbreviation or between an abbreviated title and last name. It creates a fixed (word) gap, which may be smaller than between ordinary words in case of left- and right-justified text, and it prevents an automatic word wrap or line break resp. at that position, so that the abbreviations and names like „i. e., e. g., Dr. Minor" stay together either on the old line or on the new one.

A **non-breaking hyphen** (NBH) is a hyphen, where the word-processing program will not change the lines, under no circumstances. This prevents, that in words which are combined with an abbreviation, the abbreviation stands alone on the old line. Example: „X-ray".

O

OCR (**O**ptical **C**haracter **R**ecognition) is a function when scanning. A software reads the scanned pages and converts the pixels to editable characters.

P

PDF (**P**ortable **D**ocument **F**ormat) is a page description language, which has been defined by the company Adobe. The line and page break remains unchanged, i. e. the reader sees the pages on his computer in exactly the same way as the author has created them → Acrobat Reader.

A **pictogram** is a pictorial symbol → icon.

If an image is saved in a **pixel graphics** format, the image information contains single image points (pixels). For each pixel it is saved, where it is and which color it has. Therefore the file size quickly rises, if the image area rises (opposite: → vector graphics).

PostScript (PS) is a page description language, which has been defined by the company Adobe. Nearly all current printers work with PostScript. Since Adobe take license fees for the screen display of PS files and distribute the → Acrobat Reader for PDF files free-of-charge, PDF has established as de-facto standard for the exchange of formatted documents.

Printed area is the area of a printed page, where „printing ink", i. e. texts, figures, headers and footers, tables etc. are or may be. In multi-column typesetting the printing area for each column is limited by → white space.

A **proportional font** is a font with variable character spacing (Times New Roman, Arial etc.). That means, to put it simply, that in a text written with a proportional font the distance from the end of one letter to the beginning of the next letter is constant (opposite: → fixed spacing font).

protected hyphen → non-breaking hyphen

protected space → non-breaking space

Point (pt): Unit of length in the graphical industry, e. g. used for character height and line thickness.

R

With a **rasterizing film** you can disseminate a → half tone image during copying or a negative during the enlargement into more or less coarse pixels.

Rasterizing is the dissemination of a halftone image into pixels, see above, but it is also the filling of surfaces or the background of text boxes and table cells with gray or color of different intensity. In word-processing programs this is sometimes called shading.

Reading aids are all lists/indexes and labels of a document, which exceed the pure text with figures and tables, i. e. all types of lists/indexes, footnotes, marginalia, register markings, headers and footers as well as column headings.

RFID (Radio **F**requency **Id**entification) means identification of things and creatures (e. g. containers, dogs) by means of electromagnetic waves radio chip transmits. RFID-Chips are also hidden in the spine of books in the library.

Robot → Crawler.

RTF (Rich **T**ext **F**ormat) is a file exchange format between word-processing programs, which allows the exchange of texts with formatting information to a certain extent.

S

For **scalable fonts** you can select the font size in the word-processing program. It is specified in the typographic unit → point (pt).

A **section** is a document part of the third hierarchy level in structures or tables of contents resp. In ISO 2145 this term is used on the one hand as subtopic for document parts of the third hierarchy level and on the other hand as generic term for all document parts like chapters, subchapters, sections and subsections.

Serial Item and Contribution Identifier **SICI** for articles and contributions to periodicals

Serifs are the small lines at the ends of the letters, e. g. from the Times family. While reading they help the eye to hold the line.

Small caps is a font attribute, where no small letters, but only normal-sized and smaller capital letters are used.

Soft hyphen → hyphenation proposal

Spider → Crawler.

The **structure** contains every document part number and heading, but no page numbers. It contains the logic of the contents, the "backbone". It is an intermediate result and grows with further writing of the Technical Report via the states 4-point-structure and 10-point-structure up to the final detailed structure.

The **Style Guide** is a collection of certain notations, technical terms and layout rules for a larger document (from about 20 pages on). It helps, that within a larger work the same items are always expressed (terminology) or displayed (layout) in the same way, i. e. that the work is *consistent* in itself.

T

A **table of contents (ToC)** contains for each document part the document part number, document part heading and page number and allows to quickly find chapters, subchapters, sections etc.

Terminology is the logically ordered system of terms of a field of knowledge.

In **text images** text and structuring or connecting lines/arrows/boxes are aligned image-like, so that understanding the message is facilitated and the remembering of the contents is improved. Text images are often used on slides → DTV.

Text formatting → Layout.

A **text table** is a table, which predominantly contains text.

A soft hyphen (SHY) is a **hyphenation proposal**, which is entered in he middle of a word to prevent wrong automatic hyphenation or to avoid too large distances between the words. In Word it is entered with Ctrl and „-“ in the normal keyboard. If due to text insertions or deletions the soft hyphen moves to the middle of the line it is invisible (in the printout). If you enter a normal hyphen and it moves to the middle of the line it is visible and needs to be explicitly deleted via keyboard what is often overseen or forgotten.

A **top level domain** is a name area for URLs like .de and .com. The URLs are issued by the ICANN (Internet Corporation for Assigned Names and Numbers, www.icann.org). With a whois request you can find out, whether a URL is still available and who runs an already registered URL. For the top level domains .aero, .arpa, .biz, .cat, .com, .coop, .edu, .info, .int, .jobs, .mobi, .museum, .name, .net, .org, .pro, und .travel this request is possible at www.internic.net/whois.html.

Typography is the positioning of printing ink on the paper. It is distinguished between macro-typography (on text or page level) and micro-typography (on character level).

U

A **URL** (Unified Resource Locator) is an address in the internet, e. g. the start address of a homepage like http://www.springer.com. Most office programs recognize the entry of a URL from the preceding „www.“ or „http://“ and automatically convert the address into a hyperlink.

Unicode is an international computer code for calligraphic and text symbols from all known languages, writing cultures and character systems on the earth (see also → ASCII). Unicode shall eliminate different incompatible codes in different countries or cultures. The Unicode character set is standardized in ISO 10646. One character needs 21 bit space.

V

If an image is saved in a **vector graphics** format, the image information contains scalable geometry information (e. g. center point coordinates and radius of a circle). For these geometry objects also line and fill color, line type, filling pattern etc. are saved. The file size of vector graphics is much smaller than the file size of a → pixel graphic showing the same items.

A **viewer** is a program to look at text and graphics files.

W

A **webmaster** creates and updates internet and/or intranet pages.

White space is a white area on the page, where there are no alphanumerical symbols, e. g. the spare line between two paragraphs or the white space between table cells and rows (if the cells are not limited by lines).

WIPO, the World Intellectual Property Organization, protects copyright and intellectual properties (patents, trademarks and design patents) worldwide (www.wipo.int).

The **World Wide Web Consortium** (abbreviated: **W3C**) is the board for the standardization of the technology used in the internet (www.w3.org).

WTO, the World Trade Organization, controls international trade and economy since 1995 and has their headquarters in Genf (www.wto.org).

Z

Zip file → compression.

C Index

Lightning Source UK Ltd.
Milton Keynes UK
UKOW051003220313

208010UK00003B/108/P